Soft Computing and Fractal Theory
for Intelligent Manufacturing

Studies in Fuzziness and Soft Computing

Editor-in-chief
Prof. Janusz Kacprzyk
Systems Research Institute
Polish Academy of Sciences
ul. Newelska 6
01-447 Warsaw, Poland
E-mail: kacprzyk@ibspan.waw.pl
http://www.springer.de/cgi-bin/search_book.pl?series=2941

Further volumes of this series can be found at our homepage.

Vol. 96. M. Schmitt, H.-N. Teodorescu, A. Jain, A. Jain, S. Jain and L.C. Jain (Eds.)
Computational Intelligence Processing in Medical Diagnosis, 2002
ISBN 3-7908-1463-6

Vol. 97. T. Calvo, G. Mayor and R. Mesiar (Eds.)
Aggregation Operators, 2002
ISBN 3-7908-1468-7

Vol. 98. L.C. Jain, Z. Chen and N. Ichalkaranje (Eds.)
Intelligent Agents and Their Applications, 2002
ISBN 3-7908-1469-5

Vol. 99. C. Huang and Y. Shi
Towards Efficient Fuzzy Information Processing, 2002
ISBN 3-7908-1475-X

Vol. 100. S.-H. Chen (Ed.)
Evolutionary Computation in Economics and Finance, 2002
ISBN 3-7908-1476-8

Vol. 101. S.J. Ovaska and L.M. Sztandera (Eds.)
Soft Computing in Industrial Electronics, 2002
ISBN 3-7908-1477-6

Vol. 102. B. Liu
Theory and Practice of Uncertain Programming, 2002
ISBN 3-7908-1490-3

Vol. 103. N. Barnes and Z.-Q. Liu
Knowledge-Based Vision-Guided Robots, 2002
ISBN 3-7908-1494-6

Vol. 104. F. Rothlauf
Representations for Genetic and Evolutionary Algorithms, 2002
ISBN 3-7908-1496-2

Vol. 105. J. Segovia, P.S. Szczepaniak and M. Niedzwiedzinski (Eds.)
E-Commerce and Intelligent Methods, 2002
ISBN 3-7908-1499-7

Vol. 106. P. Matsakis and L.M. Sztandera (Eds.)
Applying Soft Computing in Defining Spatial Relations, 2002
ISBN 3-7908-1504-7

Vol. 107. V. Dimitrov and B. Hodge
Social Fuzziology, 2002
ISBN 3-7908-1506-3

Vol. 108. L.M. Sztandera and C. Pastore (Eds.)
Soft Computing in Textile Sciences, 2003
ISBN 3-7908-1512-8

Vol. 109. R.J. Duro, J. Santos and M. Graña (Eds.)
Biologically Inspired Robot Behavior Engineering, 2003
ISBN 3-7908-1513-6

Vol. 110. E. Fink
Changes of Problem Representation, 2003
ISBN 3-7908-1523-3

Vol. 111. P.S. Szczepaniak, J. Segovia, J. Kacprzyk and L.A. Zadeh (Eds.)
Intelligent Exploration of the Web, 2003
ISBN 3-7908-1529-2

Vol. 112. Y. Jin
Advanced Fuzzy Systems Design and Applications, 2003
ISBN 3-7908-1537-3

Vol. 113. A. Abraham, L.C. Jain and J. Kacprzyk (Eds.)
Recent Advances in Intelligent Paradigms and Applications, 2003
ISBN 3-7908-1538-1

Vol. 114. M. Fitting and E. Orłowska (Eds.)
Beyond Two: Theory and Applications of Multiple Valued Logic, 2003
ISBN 3-7908-1541-1

Vol. 115. J.J. Buckley
Fuzzy Probabilities, 2003
ISBN 3-7908-1542-X

Vol. 116. C. Zhou, D. Maravall and D. Ruan (Eds.)
Autonomous Robotic Systems, 2003
ISBN 3-7908-1546-2

Oscar Castillo
Patricia Melin

Soft Computing and Fractal Theory for Intelligent Manufacturing

With 191 Figures
and 10 Tables

Physica-Verlag
A Springer-Verlag Company

Prof. Dr. Oscar Castillo
Prof. Dr. Patricia Melin
Tijuana Institute of Technology
Department of Computer Science
P.O. Box 4207
Chula Vista, CA 91909
USA

Av. ITR Aguascalientes 2001-A
Fracc Otay Villa real C.P. 22500
Tijuana, B.C.
Mexico
ocastillo@tectijuana.mx
pmelin@tectijuana.mx

ISBN 978-3-662-00296-4 ISBN 978-3-7908-1766-9 (eBook)
DOI 10.1007/978-3-7908-1766-9

Cataloging-in-Publication Data applied for
A catalog record for this book is available from the Library of Congress.
Bibliographic information published by Die Deutsche Bibliothek
Die Deutsche Bibliothek lists this publication in the Deutsche Nationalbibliografie; detailed bibliographic data is available in the Internet at <http://dnb.ddb.de>.

Physica-Verlag Heidelberg New York
a member of BertelsmannSpringer Science+Business Media GmbH

Softcover reprint of the hardcover 1st edition 2003

SPIN 10894231 88/3130-5 4 3 2 1 0 – Printed on acid-free paper

Preface

We describe in this book, new methods for intelligent manufacturing using soft computing techniques and fractal theory. Soft Computing (SC) consists of several computing paradigms, including fuzzy logic, neural networks, and genetic algorithms, which can be used to produce powerful hybrid intelligent systems. Fractal theory provides us with the mathematical tools to understand the geometrical complexity of natural objects and can be used for identification and modeling purposes. Combining SC techniques with fractal theory, we can take advantage of the "intelligence" provided by the computer methods and also take advantage of the descriptive power of the fractal mathematical tools. Industrial manufacturing systems can be considered as non-linear dynamical systems, and as a consequence can have highly complex dynamic behaviors. For this reason, the need for computational intelligence in these manufacturing systems has now been well recognized. We consider in this book the concept of "intelligent manufacturing" as the application of soft computing techniques and fractal theory for achieving the goals of manufacturing, which are production planning and control, monitoring and diagnosis of faults, and automated quality control.

As a prelude, we provide a brief overview of the existing methodologies in Soft Computing. We then describe our own approach in dealing with the problems in achieving intelligent manufacturing. Our particular point of view is that to really achieve intelligent manufacturing in real-world applications we need to use SC techniques and fractal theory. As consequence, we will describe several real-world applications, in which the reader will be able to appreciate that the use of these techniques really helps in achieving the goals of intelligent manufacturing. In these applications, we will always compare with the traditional approaches to make clear the advantages of using SC techniques and fractal theory.

This book is intended to be a major reference for scientists and engineers interested in applying new computational and mathematical tools to achieve intelligent manufacturing. This book can also be used as a textbook or major reference for graduate courses like the following: soft computing, intelligent manufacturing, computer-integrated manufacturing, applied artificial intelligence, and similar ones. We consider that this book can also be used to get novel ideas for new lines of research, or to continue the lines of research proposed by the authors of the book.

In Chapter one, we begin by giving a brief introduction to the main problems in achieving intelligent manufacturing in industrial plants. We discuss the importance of the concept of intelligent manufacturing. We motivate the need for using SC techniques and fractal theory for solving problems of production planning and control, monitoring and diagnosis, and quality control. We also outline the real-world applications to be considered in the book.

We describe in Chapter 2 the main ideas underlying type-1 fuzzy logic, and the application of this powerful computational theory to the problems of modeling and control. We discuss in some detail type-1 fuzzy set theory, fuzzy reasoning, and fuzzy inference systems. At the end, we also give some general guidelines for the process of fuzzy modeling. We illustrate these concepts with several examples that show the applicability of type-1 fuzzy logic. The importance of type-1 fuzzy logic as a basis for developing intelligent systems in manufacturing has been recognized in several areas of application. For this reason, we consider this chapter essential to understand the new methods for intelligent manufacturing that are described in subsequent chapters.

We describe in Chapter 3 the basic concepts, notation, and theory of type-2 fuzzy logic, which is a generalization of type-1 fuzzy logic. Type-2 fuzzy logic enables the management of uncertainty in a more complete way. This is due to the fact that in type-2 membership functions we also consider that there is uncertainty in the form of the functions, unlike type-1 membership functions in which the functions are considered to be fixed and not uncertain. We describe type-2 fuzzy set theory, type-2 fuzzy reasoning, and type-2 fuzzy systems. We also give examples to illustrate these ideas to the reader of the book.

We describe in Chapter 4 the basic concepts, notation and the learning algorithms for supervised neural networks. We discuss in some detail feed-forward neural networks, radial basis neural networks, modular neural networks, and adaptive neuro-fuzzy inference systems. First, we give a brief review of the basic concepts of neural networks and the back-propagation learning algorithm. We then continue with a general description of radial basis neural networks, and modular networks. Finally, we end the chapter with a description of the adaptive neuro-fuzzy inference system (ANFIS) method and some examples of application. The importance of supervised neural networks as a computational tool to achieve "intelligence" for software systems has been well recognized in the literature of the area. For this reason, supervised neural networks have been applied for solving complex problems of modeling, identification, and control.

We describe in Chapter 5 the basic concepts, notation and learning algorithms for unsupervised neural networks. This type of neural network only receives input data and not output data, unlike supervised neural networks, which receive input-output training data. We describe in some detail competitive neural networks, Kohonen self-organizing maps, Learning Vector Quantization (LVQ) neural networks, and Hopfield neural networks. We describe each of this type of neural networks and give examples to illustrate their applicability. Unsupervised neural networks are very important for classification, pattern recognition and clustering applications. For this reason, we consider this chapter very important for understanding some of the applications that are described in later chapters of the book.

We describe in Chapter 6 the basic concepts and notation of genetic algorithms, and simulated annealing. We also describe the application of genetic algorithms for evolving neural networks, and fuzzy systems. Both genetic algorithms and simulated annealing are basic search methodologies that can be used for system optimization. Since both techniques can be considered as general-purpose optimization methodologies, we can use any of them to find the model, which minimizes the fitting error for a specific data set. As genetic algorithms are based on the ideas of natural evolution, we can use this methodology to evolve a neural network or a fuzzy system for a particular application. The problem of finding the best architecture of a neural network is very important because there are no theoretical results on this, and in many cases we are forced to trial and error unless we use a genetic algorithm to automate this process. A similar thing occurs in finding out the optimal number of rules and membership functions of a fuzzy system for a particular application, here a genetic algorithm can also help us avoid time consuming trial and error.

We describe in Chapter 7 the basic concepts and notation of dynamical systems and fractal theory. We also describe methods for controlling chaotic behavior in non-linear systems. First, we describe the concept of a dynamical system and several methods for characterizing the different behaviors of these systems. Second, we introduce fractal theory, in particular the concept of the fractal dimension, which can be used to measure the geometrical complexity of arbitrary objects. In particular, the fractal dimension can be used to characterize the attractors of a dynamical system. Third, we describe several methods for controlling chaotic behavior in non-linear dynamical systems. In all of these cases, we illustrate the concepts and methods with examples. In this way the reader can appreciate the applicability of these concepts and methods.

We describe in Chapter 8 the application of fuzzy logic and fractal theory for solving the problems of monitoring and diagnostics in non-linear dynamic plants. In this case, we describe a hybrid approach combining fuzzy logic and fractal theory for monitoring and diagnosis, and we illustrate the advantages of the new approach with real-world examples. In the new hybrid fuzzy-fractal approach, fuzzy logic is used to represent expert knowledge on monitoring and diagnosis, and the fractal dimension is used to measure the complexity of time series of the relevant variables. We also compare the results of the fuzzy-fractal approach with conventional approaches for monitoring and diagnosis.

We describe in Chapter 9 the basic concepts and theory of adaptive model-based control of non-linear dynamic plants. We also extend the concept of adaptive control to include the use of fuzzy logic. We illustrate these concepts with the case of controlling a stepping motor drive. In this case, intelligent control of the stepping motor is achieved by using a neuro-fuzzy approach. The reason for combining neural networks with fuzzy logic was to facilitate the design of the fuzzy rule base for this application. The neural network allows the use of training data to adjust the fuzzy system for the specific application. The results of the neuro-fuzzy approach are far better than the results obtained by traditional approaches.

We describe in Chapter 10 the application of soft computing techniques and fractal theory for solving the problem of automating the quality control in sound speaker manufacturing. In this case, the problem is of analyzing the sound signals of the manufactured speakers to decide on their final quality. We use the fractal dimension to analyze the complexity of the sound signals, in this way obtaining a classification on the quality of the speakers based on the geometrical form of their signals. We also use a neuro-fuzzy approach to design a fuzzy rule base for deciding on the final quality of the manufactured speakers. The fuzzy rule base represents the human expert knowledge on quality evaluation. We compared the results of the neuro-fuzzy-fractal approach with the traditional manual approach, and of course, the results of the hybrid intelligent approach are far better than the traditional manual approach.

We describe in Chapter 11 the application of soft computing techniques to the problem of automating the electrical tuning process in the manufacturing of televisions in a real plant. In this case, the problem is of controlling the electrical tuning process of the televisions in such a way as to obtain the best quality of the image. Of course, we also have to achieve the best tuning possible in a certain amount of time to be able to produce the optimum number of televisions. For this application, we have designed a fuzzy rule base for controlling the electrical tuning process during the manufacturing of televisions. We have also used a specific genetic algorithm for optimizing the parameters of the fuzzy system. The results of automating the electrical tuning process in this manufacturing system are outstanding. Previously, human operators did perform the tuning manually and was time consuming and produced many errors.

Finally, in Chapter 12 we describe the application of soft computing techniques to the problem of controlling the charging process in the manufacturing of batteries in a real plant. We also describe the use of fuzzy logic for automating the quality control of the manufactured batteries. In this case, the first problem consists in controlling the current intensity during the charging process for the batteries, which is called "battery formation". We need to control the current intensity during the charging process in such a way as to produce the battery in the least amount of time, but without surpassing a safe temperature value. The final fuzzy controller for this charging process is obtained by using a hybrid neuro-fuzzy-genetic approach, which uses a neural network to model the process and a genetic algorithm to optimize the parameters of the fuzzy system. We did make a hardware implementation of the final fuzzy controller to really achieve the automation needed in the plant. On the other hand, we also designed a fuzzy system for automating the quality control of the manufactured batteries. This fuzzy system for quality control was also implemented in hardware by using a specific micro-controller. The results of both implementations are excellent because the accuracy and efficiency was increased with respect to the traditional manual approach used before.

We end this preface of the book by giving thanks to all the people who have helped or encourage us during the writing of this book. First of all, we would like to thank our colleague and friend Professor Janusz Kacprzyk for always supporting our work, and for motivating us to write our research work. We would also like to thank our colleagues working in Soft Computing, which are too many to mention each by their name. Of course, we need to thank our supporting agencies, CONACYT and COSNET, in our country for their help during this project. We have to thank our institution, Tijuana Institute of Technology, for always supporting our projects. Finally, we thank our families for their continuous support during the time that we spend in this project.

September, 2002 Oscar Castillo and Patricia Melin
Tijuana, Mexico

Contents

Chapter 1

Introduction

We describe in this book, new methods for intelligent manufacturing using soft computing techniques and fractal theory. Soft Computing (SC) consists of several computing paradigms, including fuzzy logic, neural networks, and genetic algorithms, which can be used to create powerful hybrid intelligent systems. Fractal theory provides us with the mathematical tools to analyze the geometrical complexity of natural and artificial objects, and can be used for identification and modeling purposes. Combining SC techniques with fractal theory, we can take advantage of the "intelligence" provided by the computer methods and also take advantage of the descriptive power of fractal mathematical tools. We consider in this book "intelligent manufacturing" as the use of SC techniques to solve manufacturing problems in industrial plants. The basic manufacturing problems that we are considering in this book are the problems of controlling the process of production, monitoring and diagnosis faults, and performing quality control. These manufacturing problems are not easy to solve because, in general, real world plants are non-linear dynamical systems, and as a consequence there is no simple way to predict their behavior. For this reason, SC techniques, which are non-linear by nature, can be used to solve these manufacturing problems.

As a prelude, we provide a brief overview of the existing methodologies for solving manufacturing problems. We then describe our own approach in dealing with these problems. Our particular point of view is that process control, monitoring and diagnosis, and quality control are problems that can not be considered apart because they are intrinsically related in real-world applications. We show in this book that process control in non-linear plants can be achieved by using fuzzy logic and/or neural networks. Monitoring and diagnosis can also be achieved by applying fuzzy logic and fractal theory. Finally, automated quality control can be achieved by applying neural networks, fuzzy logic and fractal theory. We will illustrate in this book each of these manufacturing problems and its solutions in real world situations. In each application of the SC techniques to solve a real-world manufacturing problem, we show that the intelligent approach proves to be more efficient and accurate that traditional approaches.

Traditionally, the manufacturing problems mentioned above, have been solved by using classical linear methods and models, which lack the accuracy and efficiency needed in real-world applications. Traditional methods include the use of linear statistical models and simple information systems. We instead, consider more general modeling methods, which include fuzzy logic and neural networks. We also use genetic algorithms for the optimization of the fuzzy systems and neural networks. On the other hand, we use the concept of the fractal dimension to measure the complexity of geometrical objects, which is needed for pattern recognition and time series analysis. A proper combination of these methodologies will result in a hybrid intelligent system that will solve efficiently and accurately a specific manufacturing problem.

Fuzzy logic is an area of soft computing that enables a computer system to reason with uncertainty (Castillo & Melin, 2001). A fuzzy inference system consists of a set of if-then rules defined over fuzzy sets. Fuzzy sets generalize the concept of a traditional set by allowing the membership degree to be any value between 0 and 1 (Zadeh, 1965). This corresponds, in the real world, to many situations where it is difficult to decide in an unambiguous manner if something belongs or not to a specific class. Fuzzy expert systems, for example, have been applied with some success to problems of control, diagnosis and classification, just because they can manage the complex expert reasoning involved in these areas of application. The main disadvantage of fuzzy systems is that they can't adapt to changing situations. For this reason, it is a good idea to combine fuzzy logic with neural networks or genetic algorithms, because either one of these last two methodologies could give adaptability to the fuzzy system (Melin & Castillo, 2002). On the other hand, the knowledge that is used to build these fuzzy rules is uncertain. Such uncertainty leads to rules whose antecedents or consequents are uncertain, which translates into uncertain antecedent or consequent membership functions (Karnik & Mendel 1998). Type-1 fuzzy systems, like the ones mentioned above, whose membership functions are type-1 fuzzy sets, are unable to directly handle such uncertainties. We also describe in this book, type-2 fuzzy systems, in which the antecedent or consequent membership functions are type-2 fuzzy sets. Such sets are fuzzy sets whose membership grades themselves are type-1 fuzzy sets; they are very useful in circumstances where it is difficult to determine an exact membership function for a fuzzy set.

Neural networks are computational models with learning (or adaptive) characteristics that model the human brain (Jang, Sun & Mizutani, 1997). Generally speaking, biological natural neural networks consist of neurons and connections between them, and this is modeled by a graph with nodes and arcs to form the computational neural network. This graph along with a computational algorithm to specify the learning capabilities of the system is what makes the neural network a powerful methodology to simulate intelligent or expert behavior (Miller, Sutton & Werbos, 1995). Neural networks can be classified in supervised and unsupervised. The main difference is that in the case of the supervised neural networks the learning algorithm uses input-output training data to model the dynamic system, on the other hand, in the case of unsupervised neural networks only the input data is given. In the case of an unsupervised network, the input data is used to make representative clusters of all the data. It has been shown, that neural networks are universal approximators, in the sense that they can model any general function to a specified accuracy and for this reason neural networks have been applied to problems of system identification, control, diagnosis, and time series prediction.

Genetic algorithms and simulated annealing are optimization methodologies based on principles of nature (Jang, Sun & Mizutani, 1997). Both methodologies can also be viewed as searching algorithms because they explore a space using heuristics inspired by nature. Genetic algorithms are based on the ideas of evolution and the biological process that occur at the DNA level. Basically, a genetic algorithm uses a population of individuals, which are modified by using genetic operators in such a way as to eventually obtain the fittest individual (Man, Tang & Kwong, 1999). Any optimization problem has to be represented by using chromosomes, which are a codified representation of the real values of the variables in the problem (Mitchell, 1998). On the other hand, simulated annealing is based on the thermodynamics of the cooling process in metals. The cooling process in metals has to be done in a controlled way to obtain the desired characteristics in the metal. In this case, the search algorithm uses an energy function, which is decreased in each step according to a cooling schedule. Both, genetic algorithms and simulated annealing can be used to optimize a general function. As consequence, one of the main applications of both methodologies is in the optimization of neural networks and fuzzy systems for specific applications.

Fractal theory is the study of the basic concepts of fractals, fractal geometry and fractal dimension. Fractal geometry is a mathematical tool for dealing with complex systems that have no characteristic length scale (Mandelbrot, 1987). A well-known example is the shape of a coastline. When we see two pictures of a coastline on two different scales, we can't tell which scale belongs to which picture: both look the same. This means that the coastline is scale-invariant or, equivalently, has no characteristic length scale. Other examples in nature are rivers, cracks, mountains, and clouds. Scale-invariant systems, are usually characterized by non-integer "fractal" dimensions (Melin & Castillo, 2002). In our case, for manufacturing applications the use of the fractal dimension is as measure of complexity of signals, images or time series. For example, in the case of monitoring and diagnosis, we can use the time series of the relevant variables of the process to identify faults or problems. Another application is when we determine the quality of a product base on certain geometrical characteristics that it posses. In any case, the use of the fractal dimension can help in certain applications where the complexity needs to be analyzed to solve the problems (Castillo & Melin, 2001). Of course, when we combine the use of the fractal dimension with SC techniques, we will be constructing a hybrid intelligent system for a specific manufacturing application.

The diversity of the manufacturing applications considered in this work, gives an idea of the universality of the hybrid approaches presented in this book. The hybrid approaches for achieving intelligent manufacturing combine the use of SC techniques with fractal theory. The best combination of SC techniques with the fractal dimension for a specific application may change because of the properties of the system under consideration, but one can always find the hybrid architecture needed for achieving the ultimate goal of intelligent manufacturing. Of course, we still need to do a lot of work in finding out general rules for determining in advance the best combination of techniques for a specific application. The best architecture for intelligent manufacturing applications has to be determined in many cases by a lot of experimental work or by using an evolutionary approach for evolving the design of the intelligent system.

Chapter 2

Type-1 Fuzzy Logic

This chapter introduces the basic concepts, notation, and basic operations for the type-1 fuzzy sets that will be needed in the following chapters. Type-2 fuzzy sets as well as their operations will be discussed in the next chapter. For this reason, in this chapter we will focus only on type-1 fuzzy logic. Since research on fuzzy set theory has been underway for over 30 years now, it is practically impossible to cover all aspects of current developments in this area. Therefore, the main goal of this chapter is to provide an introduction to and a summary of the basic concepts and operations that are relevant to the study of type-1 fuzzy sets. We also introduce in this chapter the definition of linguistic variables and linguistic values and explain how to use them in type-1 fuzzy rules, which are an efficient tool for quantitative modelling of words or sentences in a natural or artificial language. By interpreting fuzzy rules as fuzzy relations, we describe different schemes of fuzzy reasoning, where inference procedures based on the concept of the compositional rule of inference are used to derive conclusions from a set of fuzzy rules and known facts. Fuzzy rules and fuzzy reasoning are the basic components of fuzzy inference systems, which are the most important modelling tool, based on fuzzy set theory.

The "fuzzy inference system" is a popular computing framework based on the concepts of fuzzy set theory, fuzzy if-then rules, and fuzzy reasoning (Jang, Sun & Mizutani, 1997). It has found successful applications in a wide variety of fields, such as automatic control, data classification, decision analysis, expert systems, time series prediction, robotics, and pattern recognition (Jamshidi, 1997). Because of its multidisciplinary nature, the fuzzy inference system is known by numerous other names, such as "fuzzy expert system" (Kandel, 1992), "fuzzy model" (Sugeno & Kang, 1988), "fuzzy associative memory" (Kosko, 1992), and simply "fuzzy system".

The basic structure of a type-1 fuzzy inference system consists of three conceptual components: a "rule base", which contains a selection of fuzzy rules; a "data base" (or "dictionary"), which defines the membership functions used in the fuzzy rules; and a "reasoning mechanism", which performs the inference procedure upon the rules and given facts to derive a reasonable output or conclusion. In general, we can say that a fuzzy inference system implements a non-linear mapping from its input space to output space. This mapping is accomplished by a number of fuzzy if-then rules, each of which describes the local behavior of the mapping. In particular, the antecedent of a rule defines a fuzzy region in the input space, while the consequent specifies the output in the fuzzy region.

We will describe in the following chapter a new area in fuzzy logic, which studies type-2 fuzzy sets and type-2 fuzzy systems. Basically, a type-2 fuzzy set is a set in which we also have uncertainty about the membership function. Since we are dealing with uncertainty for the conventional fuzzy sets (which are called type-1 fuzzy sets here) we can achieve a higher degree of approximation in modelling real world problems. Of course, type-2 fuzzy systems consist of fuzzy if-then rules, which contain type-2 fuzzy sets. We can say that type-2 fuzzy logic is a generalization of conventional fuzzy logic (type-1) in the sense that uncertainty is not only limited to the linguistic variables but also is present in the definition of the membership functions.

In what follows, we shall first introduce the basic concepts of fuzzy sets, and fuzzy reasoning. Then we will introduce and compare the three types of fuzzy inference systems that have been employed in various applications. Finally, we will address briefly the features and problems of fuzzy modelling, which is concerned with the construction of fuzzy inference systems for modelling a given target system. In this chapter, we will assume that all fuzzy sets, fuzzy rules and operations are of type-1 category, unless otherwise specified.

2.1 Type-1 Fuzzy Set Theory

Let X be a space of objects and x be a generic element of X. A classical set A, $A \subseteq X$, is defined by a collection of elements or objects $x \in X$, such that each x can either belong or not belong to the set A. By defining a "characteristic function" for each element $x \in X$, we can represent a classical set A by a set of order pairs (x,0) or (x,1), which indicates $x \notin A$ or $x \in A$, respectively.

Unlike the aforementioned conventional set, a fuzzy set (Zadeh, 1965) expresses the degree to which an element belong to a set. Hence the characteristic function of a fuzzy set is allowed to have values between 0 and 1, which denotes the degree of membership of an element in a given set.

Definition 2.1 **Fuzzy sets and membership functions**

If X is a collection of objects denoted generically by x, then a "fuzzy set" A in X is defined as a set of ordered pairs:

$$A = \{(x, \mu_A(x)) \mid x \in X\}. \qquad (2.1)$$

where $\mu_A(x)$ is called "membership function" (or MF for short) for the fuzzy set A. The MF maps each element of X to a membership grade (or membership value) between 0 and 1.

Obviously, the definition of a fuzzy set is a simple extension of the definition of a classical set in which the characteristic function is permitted to have any values between 0 and 1. If the values of the membership function $\mu_A(x)$ is restricted to either 0 or 1, then A is reduced to a classical set and $\mu_A(x)$ is the characteristic function of A. This can be seen with the following example.

Example 2.1 *Fuzzy set with a discrete universe of discourse X*
Let X = {Tijuana, Monterrey, Cancun} be the set of cities one may choose to organize a conference in. The fuzzy set A = "desirable city to organize a conference in" may be described as follows:

A = {(Tijuana, 0.5), (Monterrey, 0.7), (Cancun, 0.9)}

In this case, the universe of discourse X is discrete- in this example, three cities in Mexico. Of course, the membership grades listed above are quite subjective; anyone can come up with three different values according to his or her preference.

A fuzzy set is uniquely specified by its membership function. To describe membership functions more specifically, we shall define the nomenclature used in the literature (Jang, Sun & Mizutani, 1997).

Definition 2.2 **Support**
The "support" of a fuzzy set A is the set of all points x in X such that $\mu_A(x) > 0$:

$$\text{support}\ (A) = \{\ x \mid \mu_A(x) > 0\ \}. \tag{2.2}$$

Definition 2.3 **Core**
The "core" of a fuzzy set is the set of all points x in X such that $\mu_A(x) = 1$:

$$\text{core}\ (A) = \{\ x \mid \mu_A(x) = 1\ \}. \tag{2.3}$$

Definition 2.4 **Normality**
A fuzzy set A is "normal" if its core is nonempty. In other words, we can always find a point $x \in X$ such that $\mu_A(x) = 1$.

Definition 2.5 **Crossover points**
A "crossover point" of a fuzzy set A is a point $x \in X$ at which $\mu_A(x) = 0.5$:

$$\text{crossover}\ (A) = \{\ x \mid \mu_A(x) = 0.5\ \}. \tag{2.4}$$

Definition 2.6 **Fuzzy singleton**
A fuzzy set whose support is a single point in X with $\mu_A(x) = 1$ is called a "fuzzy singleton".

Definition 2.7 **α-cut, strong α-cut**
The "α-cut" or "α-level set" of a fuzzy set A is a crisp set defined by

$$A_\alpha = \{ x \mid \mu_A(x) \geq \alpha \}. \tag{2.5}$$

"Strong α-cut" or "strong α–level set" are defined similarly:

$$A'_\alpha = \{ x \mid \mu_A(x) > \alpha \}. \tag{2.6}$$

Using the notation for a level set, we can express the support and core of a fuzzy set A as

$$\text{support}(A) = A'_0$$

and

$$\text{core}(A) = A_1$$

respectively.

Corresponding to the ordinary set operations of union, intersection and complement, fuzzy sets have similar operations, which were initially defined in Zadeh's seminal paper (Zadeh, 1965). Before introducing these three fuzzy set operations, first we shall define the notion of containment, which plays a central role in both ordinary and fuzzy sets. This definition of containment is, of course, a natural extension of the case for ordinary sets.

Definition 2.8 **Containment**
The fuzzy set A is "contained" in fuzzy set B (or, equivalently, A is a "subset" of B) if and only if $\mu_A(x) \leq \mu_B(x)$ for all x. Mathematically,

$$A \subseteq B \Leftrightarrow \mu_A(x) \leq \mu_B(x). \tag{2.7}$$

Definition 2.9 **Union**
The "union" of two fuzzy sets A and B is a fuzzy set C, written as $C = A \cup B$ or C = A OR B, whose MF is related to those of A and B by

$$\mu_C(x) = \max(\mu_A(x), \mu_B(x)) = \mu_A(x) \vee \mu_B(x). \tag{2.8}$$

Definition 2.10 **Intersection**
The "intersection" of two fuzzy sets A and B is a fuzzy set C, written as C = A∩B or C = A AND B, whose MF is related to those of A and B by

$$\mu_C(x) = \min(\ \mu_A(x), \mu_B(x)\ ) = \mu_A(x) \wedge \mu_B(x). \quad (2.9)$$

Definition 2.11 **Complement or Negation**
The "complement" of a fuzzy set A, denoted by A (¬A, NOT A), is defined as

$$\mu_A(x) = 1 - \mu_A(x). \quad (2.10)$$

As mentioned earlier, a fuzzy set is completely characterized by its MF. Since most fuzzy sets in use have a universe of discourse X consisting of the real line R, it would be impractical to list all the pairs defining a membership function. A more convenient and concise way to define a MF is to express it as a mathematical formula. First we define several classes of parameterized MFs of one dimension.

Definition 2.12 **Triangular MFs**
A "triangular MF" is specified by three parameters {a, b, c} as follows:

$$y = \text{triangle}(x;a,b,c) = \begin{cases} 0, & x \le a. \\ (x-a)\,/\,(b-a), & a \le x \le b. \\ (c-x)\,/\,(c-b), & b \le x \le c. \\ 0, & c \le x. \end{cases} \quad (2.11)$$

The parameters {a,b,c} (with a < b < c) determine the x coordinates of the three corners of the underlying triangular MF.

Figure 2.1 (a) illustrates a triangular MF defined by triangle(x; 10, 20, 40).

Definition 2.13 **Trapezoidal MFs**
A "trapezoidal MF" is specified by four parameters {a, b, c, d} as follows:

$$\text{trapezoid}\ (x;a,b,c,d) = \begin{cases} 0, & x \le a. \\ (x-a)\,/\,(b-a), & a \le x \le b. \\ 1, & b \le x \le c. \\ (d-x)\,/\,(d-c), & c \le x \le d. \\ 0, & d \le x. \end{cases} \quad (2.12)$$

The parameters {a, b, c, d} (with $a < b \leq c < d$) determine the x coordinates of the four corners of the underlying trapezoidal MF. Figure 2.1 (b) illustrates a trapezoidal MF defined by trapezoid(x; 10, 20 40, 75).

Due to their simple formulas and computational efficiency, both triangular MFs and trapezoidal MFs have been used extensively, especially in real-time implementations. However, since the MFs are composed of straight line segments, they are not smooth at the corner points specified by the parameters. In the following we introduce other types of MFs defined by smooth and nonlinear functions.

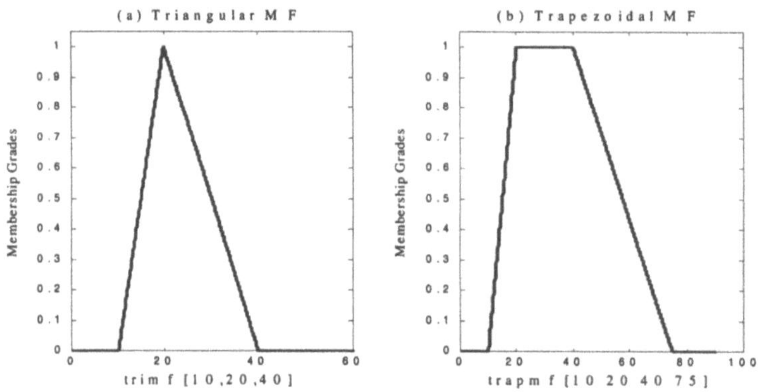

(a) Triangular MF b) Trapezoidal MF

Figure 2.1 Examples of two types of parameterized MFs

Definition 2.14 **Gaussian MFs**

A "Gaussian MF" is specified by two parameters {c , σ }:

$$\text{gaussian}(x; c, \sigma) = e^{-\frac{1}{2}\left(\frac{x-c}{\sigma}\right)^2} . \qquad (2.13)$$

A "Gaussian" MF is determined completely by c and σ ; c represents the MFs center and σ determines the MFs width. Figure 2.2 (a) plots a Gaussian MF defined by gaussian (x; 50, 20).

Definition 2.15 **Generalized bell MFs**

A "generalized bell MF" is specified by three parameters {a, b, c}:

$$\text{bell}(x; a, b, c) = \frac{1}{1 + |(x-c)/a|^{2b}} \qquad (2.14)$$

where the parameter b is usually positive. We can note that this MF is a direct generalization of the Cauchy distribution used in probability theory, so it is also referred to as the "Cauchy MF". Figure 2.2 (b) illustrates a generalized bell MF defined by bell(x; 20, 4, 50).

Although the Gaussian MFs and bell MFs achieve smoothness, they are unable to specify asymmetric MFs, which are important in certain applications. Next we define the sigmoidal MF, which is either open left or right.

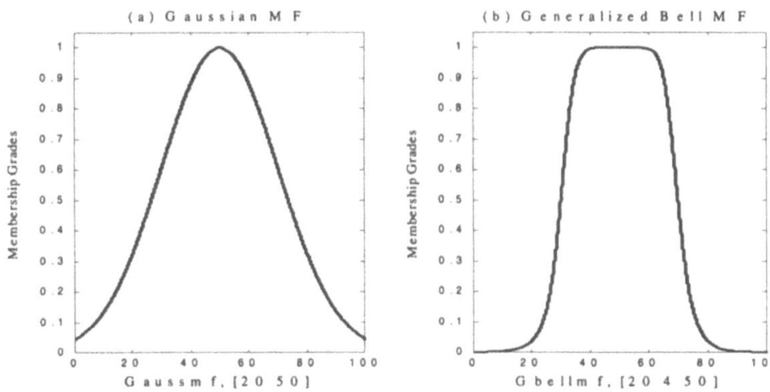

(a) Gaussian MF (b) Generalized Bell MF

Figure 2.2 Examples of two classes of parameterized continuous MFs.

<u>Definition</u> 2.16 **Sigmoidal MFs**

A "Sigmoidal MF" is defined by the following equation:

$$\text{sig}(x; a, c) = \frac{1}{1 + \exp[-a(x-c)]} \tag{2.15}$$

where a controls the slope at the crossover point x = c.

Depending on the sign of the parameter "a", a sigmoidal MF is inherently open right or left and thus is appropriate for representing concepts such as "very large" or "very negative". Figure 2.3 shows two sigmoidal functions y_1 =sig(x; 1, -5) and y_2 =sig(x; -2, 5).

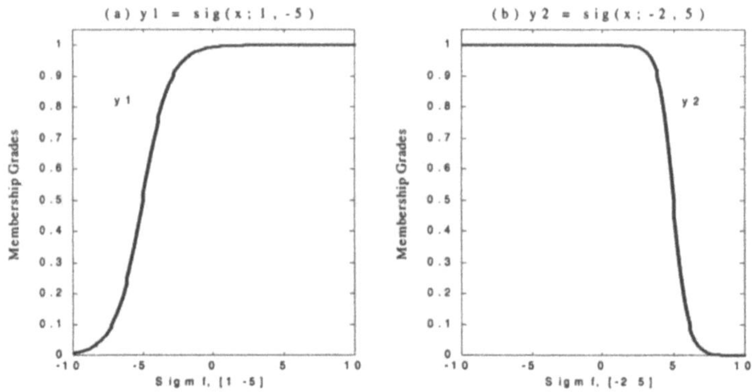

(a) $y_1 = \text{sig}(x; 1, -5)$ (b) $y_2 = \text{sig}(x; -2, 5)$

Figure 2.3 Two sigmoidal functions y_1 and y_2 .

2.2 Fuzzy Rules and Fuzzy Reasoning

In this section we introduce the concepts of the extension principle and fuzzy relations, which extend the notions of fuzzy sets introduced previously. Then we give the definition of linguistic variables and linguistic values and show how to use them in fuzzy rules. By interpreting fuzzy rules as fuzzy relations, we describe different schemes of fuzzy reasoning. Fuzzy rules and fuzzy reasoning are the backbone of fuzzy inference systems, which are the most important modeling tool based on fuzzy set theory.

2.2.1 Fuzzy Relations

The "extension principle" is a basic concept of fuzzy set theory that provides a general procedure for extending crisp domains of mathematical expressions to fuzzy domains. This procedure generalizes a common one-to-one mapping of a function f to a mapping between fuzzy sets. More specifically, lets assume that f is a function from X to Y and A is a fuzzy set on X defined as

$$A = \mu_A(x_1)/x_1 + \mu_A(x_2)/x_2 + \ldots + \mu_A(x_n)/x_n$$

Then the extension principle states that the image of fuzzy set A under the mapping f can be expressed as a fuzzy set B,

$$B = f(A) = \mu_A(x_1)/y_1 + \mu_A(x_2)/y_2 + \ldots + \mu_A(x_n)/y_n$$

where $y_i = f(x_i)$, $i = 1, \ldots, n$. In other words, the fuzzy set B can be defined through the values of f in $x_1, x_2, \ldots, x_n$. If f is a many-to-one mapping, then there exists $x_1, x_2 \in X$, $x_1 \neq x_2$, such that $f(x_1) = f(x_2) = y^*$, $y^* \in Y$. In this case, the membership grade of B at $y = y^*$ is the maximum of the membership grades of A at $x = x_1$ and $x = x_2$, since $f(x) = y^*$ may result from $x = x_1$ or $x = x_2$. More generally speaking, we have

$$\mu_B(y) = \max_{x = f^{-1}(y)} \mu_A(x).$$

A simple example of this concept is shown below.

Example 2.2 *Application of the extension principle to fuzzy sets*

Lets suppose we have the following fuzzy set with discrete universe

$$A = 0.2/-2 + 0.5/-1 + 0.7/0 + 0.9/1 + 0.4/2$$

and lets suppose that we have the following mapping

$$y = x^2 + 1.$$

After applying the extension principle, we have the following result

$$B = 0.2/5 + 0.5/2 + 0.7/1 + 0.9/2 + 0.4/5$$
$$B = 0.7/1 + (0.2 \vee 0.4)/5 + (0.5 \vee 0.9)/2$$
$$B = 0.7/1 + 0.4/5 + 0.9/2,$$

where $\vee$ represents "max".

Binary fuzzy relations are fuzzy sets in $X \times Y$ which map each element in $X \times Y$ to a membership grade between 0 and 1. In particular, unary fuzzy relations are fuzzy sets with one-dimensional MFs; binary fuzzy relations are fuzzy sets with two-dimensional MFs, and so on. Here we will restrict our attention to binary fuzzy relations. A generalization to n-ary fuzzy relations is not so difficult.

Definition 2.17 **Binary fuzzy relation**
Let X and Y be two universes of discourse. Then

$$\mathfrak{R} = \{ ((x, y), \mu_{\mathfrak{R}}(x, y)) \mid (x, y) \in X \times Y \} \tag{2.16}$$

is a binary fuzzy relation in $X \times Y$.

Example 2.3 *Binary fuzzy relations*
Let $X = \{1, 2, 3\}$ and $Y = \{1, 2, 3, 4, 5\}$ and $\Re$ = "y is slightly greater than x". The MF of the fuzzy relation $\Re$ can be defined (subjectively) as

$$\mu_{\Re}(x, y) = \begin{cases} (y - x)/(y + x), & \text{if } y > x. \\ 0, & \text{if } y \leq x. \end{cases} \tag{2.17}$$

This fuzzy relation $\Re$ can be expressed as a relation matrix in the following form:

$$\Re = \begin{pmatrix} 0 & 0.333 & 0.500 & 0.600 & 0.666 \\ 0 & 0 & 0.200 & 0.333 & 0.428 \\ 0 & 0 & 0 & 0.142 & 0.250 \end{pmatrix}$$

where the element at row i and column j is equal to the membership grade between the *i*th element of X and *j*th element of Y.

Other common examples of binary fuzzy relations are the following:

- x is similar to y (x and y are objects)
- x depends on y (x and y are events)
- If x is big, then y is small (x is an observed reading and y is the corresponding action)

The last example, "If x is A, then y is B", is used repeatedly in fuzzy systems. We will explore fuzzy relations of this type in the following section.

Fuzzy relations in different product spaces can be combined through a composition operation. Different composition operations have been proposed for fuzzy relations; the best known is the max-min composition proposed by Zadeh in 1965.

Definition 2.18 **Max-min composition**
Let $\Re_1$ and $\Re_2$ be two fuzzy relations defined on $X \times Y$ and $Y \times Z$, respectively. The "max-min composition" of $\Re_1$ and $\Re_2$ is a fuzzy set defined by

$$\Re_1 o \Re_2 = \{[(x, z), \max_y \min(\mu_{\Re 1}(x, y), \mu_{\Re 2}(y, z))] | x \in X, y \in Y, z \in Z\} \tag{2.18}$$

When $\Re_1$ and $\Re_2$ are expressed as relation matrices, the calculation of the composition $\Re_1 o \Re_2$ is almost the same as matrix multiplication, except that $\times$ and $+$ are replaced by the "min" and "max" operations, respectively. For this reason, the max-min composition is also called the "max-min product".

2.2.2 Fuzzy Rules

As was pointed out by Zadeh in his work on this area (Zadeh, 1973), conventional techniques for system analysis are intrinsically unsuited for dealing with humanistic systems, whose behavior is strongly influenced by human judgment, perception, and emotions. This is a manifestation of what might be called the "principle of incompatibility": "As the complexity of a system increases, our ability to make precise and yet significant statements about its behavior diminishes until a threshold is reached beyond which precision and significance become almost mutually exclusive characteristics" (Zadeh, 1973). It was because of this belief that Zadeh proposed the concept of linguistic variables (Zadeh, 1971) as an alternative approach to modelling human thinking.

Definition 2.19 **Linguistic variables**
A "Linguistic variable" is characterized by a quintuple (x, T(x), X, G, M) in which x is the name of the variable; T(x) is the "term set" of x-that is, the set of its "linguistic values" or "linguistic terms"; X is the universe of discourse, G is a "syntactic rule" which generates the terms in T(x); and M is a "semantic rule" which associates with each linguistic value A its meaning M(A), where M(A) denotes a fuzzy set in X.

Definition 2.20 **Concentration and dilation of linguistic values**
Let A be a linguistic value characterized by a fuzzy set membership function $\mu_A(.)$. Then A^k is interpreted as a modified version of the original linguistic value expressed as

$$A^k = \int_X [\mu_A(x)]^k / x \ . \tag{2.19}$$

In particular, the operation of "concentration" is defined as

$$\text{CON}(A) = A^2 \ , \tag{2.20}$$

while that of "dilatation" is expressed by

$$\text{DIL}(A) = A^{0.5} \ . \tag{2.21}$$

Conventionally, we take CON(A) and DIL(A) to be the results of applying the hedges "very" and "more or less", respectively, to the linguistic term A. However, other consistent definitions for these linguistic hedges are possible and well justified for various applications.

Following the definitions given before, we can interpret the negation operator NOT and the connectives AND and OR as

$$
\begin{aligned}
&\text{NOT}(A) = \neg A = \int_X [\,1 - \mu_A(x)\,] / x \quad , \\
&A \text{ AND } B = A \cap B = \int_X [\,\mu_A(x) \wedge \mu_B(x)\,] / x \, , \qquad (2.22) \\
&A \text{ OR } B = A \cup B = \int_X [\,\mu_A(x) \vee \mu_B(x)\,] / x \quad .
\end{aligned}
$$

respectively, where A and B are two linguistic values whose meanings are defined by $\mu_A(.)$ and $\mu_B(.)$.

Definition 2.21 **Fuzzy If-Then Rules**
A "fuzzy if-then rule" (also known as "fuzzy rule", "fuzzy implication", or "fuzzy conditional statement") assumes the form

$$\textbf{if } x \text{ is } A \textbf{ then } y \text{ is } B \quad , \qquad (2.23)$$

where A and B are linguistic values defined by fuzzy sets on universes of discourse X and Y, respectively. Often "x is A" is called "antecedent" or "premise", while "y is B" is called the "consequence" or "conclusion".

Examples of fuzzy if-then rules are widespread in our daily linguistic expressions, such as the following:

- If pressure is high, then volume is small.
- If the road is slippery, then driving is dangerous.
- If the speed is high, then apply the brake a little.

Before we can employ fuzzy if-then rules to model and analyze a system, first we have to formalize what is meant by the expression "if x is A then y is B", which is sometimes abbreviated as $A \rightarrow B$. In essence, the expression describes a relation between two variables x and y; this suggests that a fuzzy if-then rule is defined as a binary fuzzy relation R on the product space $X \times Y$. Generally speaking, there are two ways to interpret the fuzzy rule $A \rightarrow B$. If we interpret $A \rightarrow B$ as A "coupled with" B then

$$R = A \rightarrow B = A \times B = \int_{X \times Y} \mu_A(x) * \mu_B(y) / (x,y)$$

where * is an operator for intersection (Mamdani & Assilian, 1975). On the other hand, if $A \rightarrow B$ is interpreted as A "entails" B, then it can be written as one of two different formulas:

- Material implication:
$$R = A \rightarrow B = \neg A \cup B \ . \qquad (2.24)$$
- Propositional Calculus:
$$R = A \rightarrow B = \neg A \cup (A \cap B) \ . \qquad (2.25)$$

Although these two formulas are different in appearance, they both reduce to the familiar identity $A \rightarrow B \equiv \neg A \cup B$ when A and B are propositions in the sense of two-valued logic.

Fuzzy reasoning, also known as approximate reasoning, is an inference procedure that derives conclusions from a set of fuzzy if-then rules and known facts. The basic rule of inference in traditional two-valued logic is "modus ponens", according to which we can infer the truth of a proposition B from the truth of A and the implication $A \rightarrow B$. This concept is illustrated as follows:

premise 1 (fact):	x is A ,
premise 2 (rule):	if x is A then y is B ,
consequence (conclusion):	y is B .

However, in much of human reasoning, modus ponens is employed in an approximate manner. This is written as

premise 1 (fact):	x is A'
premise 2 (rule):	if x is A then y is B ,
consequence (conclusion):	y is B'

where A' is close to A and B' is close to B. When A, B, A' and B' are fuzzy sets of appropriate universes, the foregoing inference procedure is called "approximate reasoning" or "fuzzy reasoning"; it is also called "generalized modus ponens" (GMP for short), since it has modus ponens as a special case.

Definition 2.22 **Fuzzy reasoning**

Let A, A', and B be fuzzy sets of X, X, and Y respectively. Assume that the fuzzy implication $A \rightarrow B$ is expressed as a fuzzy relation R on $X \times Y$. Then the fuzzy set B induced by "x is A'" and the fuzzy rule "if x is A then y is B" is defined by

$$\mu_{B}'(y) = \max_X \min [\, \mu_A'(x), \mu_R(x, y) \,]$$
$$= \vee_X [\, \mu_A'(x) \wedge \mu_R(x, y) \,] . \qquad (2.26)$$

Now we can use the inference procedure of fuzzy reasoning to derive conclusions provided that the fuzzy implication $A \rightarrow B$ is defined as an appropriate binary fuzzy relation.

Single Rule with Single Antecedent

This is the simplest case, and the formula is available in Equation (2.26). A further simplification of the equation yields

$$\mu_B'(y) = [\, \vee_X (\, \mu_A'(x) \wedge \mu_A(x) \,) \,] \wedge \mu_B(y)$$

$$= \omega \wedge \mu_B(y)$$

In other words, first we find the degree of match ω as the maximum of $\mu_{A'}(x) \wedge \mu_A(x)$; then the MF of the resulting B' is equal to the MF of B clipped by ω. Intuitively, ω represents a measure of degree of belief for the antecedent part of a rule; this measure gets propagated by the if-then rules and the resulting degree of belief or MF for the consequent part should be no greater than ω.

Multiple Rules with Multiple Antecedents

The process of fuzzy reasoning or approximate reasoning for the general case can be divided into four steps:

1) Degrees of compatibility: Compare the known facts with the antecedents of fuzzy rules to find the degrees of compatibility with respect to each antecedent MF.
2) Firing Strength: Combine degrees of compatibility with respect to antecedent MFs in a rule using fuzzy AND or OR operators to form a firing strength that indicates the degree to which the antecedent part of the rule is satisfied.
3) Qualified (induced) consequent MFs: Apply the firing strength to the consequent MF of a rule to generate a qualified consequent MF.
4) Overall output MF: Aggregate all the qualified consequent MFs to obtain an overall output MF.

2.3 Fuzzy Inference Systems

In this section we describe the three types of fuzzy inference systems that have been widely used in the applications. The differences between these three fuzzy inference systems lie in the consequents of their fuzzy rules, and thus their aggregation and defuzzification procedures differ accordingly.

The "Mamdani fuzzy inference system" (Mamdani & Assilian, 1975) was proposed as the first attempt to control a steam engine and boiler combination by a set of linguistic control rules obtained from experienced human operators. Figure 2.4 is an illustration of how a two-rule Mamdani fuzzy inference system derives the overall output z when subjected to two numeric inputs x and y.

In Mamdani's application, two fuzzy inference systems were used as two controllers to generate the heat input to the boiler and throttle opening of the engine cylinder, respectively, to regulate the steam pressure in the boiler and the speed of the engine. Since the engine and boiler take only numeric values as inputs, a defuzzifier was used to convert a fuzzy set to a numeric value.

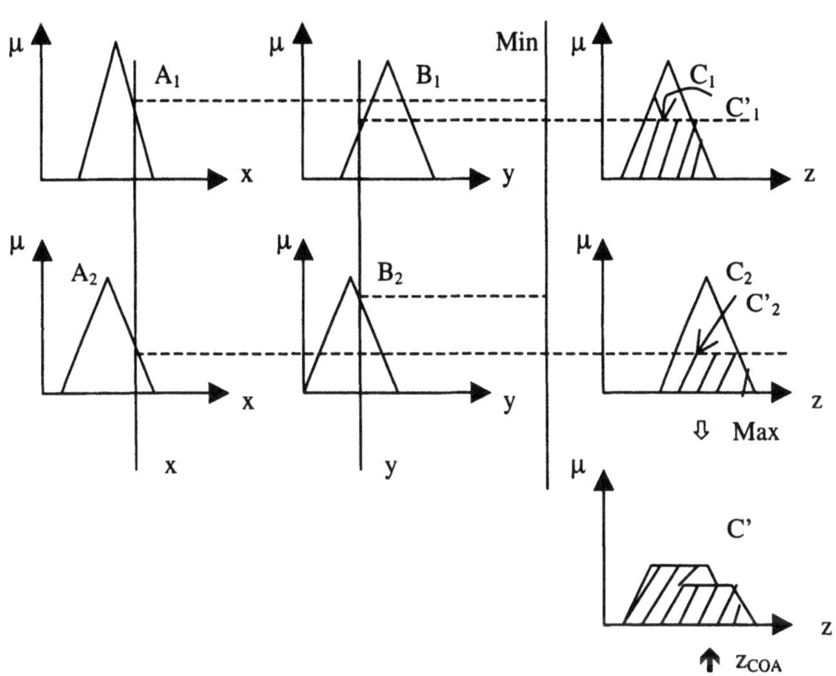

Figure 2.4 The Mamdani fuzzy inference system using the min and max operators.

Defuzzification

Defuzzification refers to the way a numeric value is extracted from a fuzzy set as a representative value. In general, there are five methods for defuzzifying a fuzzy set A of a universe of discourse Z, as shown in Figure 2.5 (Here the fuzzy set A is usually represented by an aggregated output MF, such as C' in Figure 2.4). A brief explanation of each defuzzification strategy follows.

- Centroid of area z_{COA}:

$$z_{COA} = \frac{\int_Z \mu_A(z) z dz}{\int_Z \mu_A(z) dz} \tag{2.27}$$

where $\mu_A(z)$ is the aggregated output MF. This is the most widely adopted defuzzification strategy, which is reminiscent of the calculation of expected values of probability distributions.

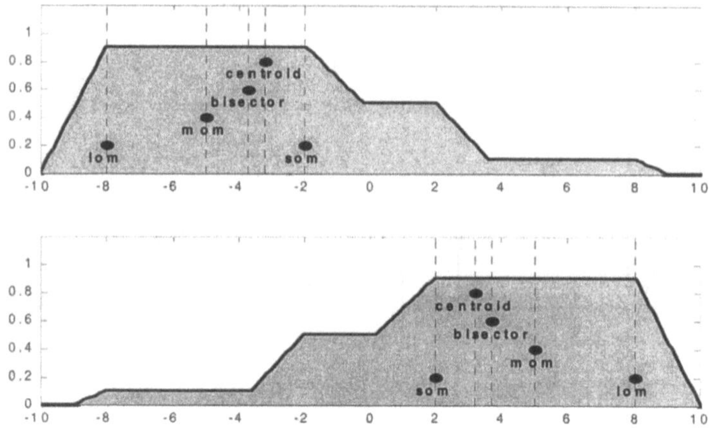

Figure 2.5 Various defuzzification methods for obtaining a numeric output.

- Bisector of area z_{BOA} : z_{BOA} satisfies

$$\int_{\alpha}^{z_{BOA}} \mu_A(z)dz = \int_{z_{BOA}}^{\beta} \mu_A(z)dz \quad , \tag{2.28}$$

where $\alpha = \min\{z \mid z \in Z\}$ and $\beta = \max\{z \mid z \in Z\}$.

- Mean of maximum z_{MOM} : z_{MOM} is the average of the maximizing z at which the MF reach a maximum μ^*. Mathematically,

$$z_{MOM} = \frac{\int_{z'} z dz}{\int_{z'} dz} \quad , \tag{2.29}$$

where $z' = \{ z \mid \mu_A(z) = \mu^* \}$. In particular, if $\mu_A(z)$ has a single maximum at $z = z^*$, then $z_{MOM} = z^*$. Moreover, if $\mu_A(z)$ reaches its maximum whenever $z \in [z_{left}, z_{right}]$ then $z_{MOM} = (z_{left} + z_{right}) / 2$.

- Smallest of maximum z_{SOM} : z_{SOM} is the minimum (in terms of magnitude) of the maximizing z.
- Largest of maximum z_{LOM} : z_{LOM} is the maximum (in terms of magnitude) of the maximizing z. Because of their obvious bias, z_{SOM}

and z_{LOM} are not used as often as the other three defuzzification methods.

The calculation needed to carry out any of these five defuzzification operations is time-consuming unless special hardware support is available. Furthermore, these defuzzification operations are not easily subject to rigorous mathematical analysis, so most of the studies are based on experimental results. This leads to the propositions of other types of fuzzy inference systems that do not need defuzzification at all; two of them will be described in the following. Other more flexible defuzzification methods can be found in several more recent papers (Yager & Filev, 1993), (Runkler & Glesner, 1994).

We will give a simple example to illustrate the use of the Mamdani fuzzy inference system. We will consider the case of determining the quality of a manufactured material based on the input variables: roughness and porosity. The fuzzy model will consist of a set of rules relating these variables. In Figure 2.6 we show the architecture of the fuzzy system, which was obtained by using the MATLAB Fuzzy Logic Toolbox. We show in Figure 2.7 the fuzzy rule base, which was obtained by using the "rule editor" of the same toolbox. In Figure 2.8 we can appreciate the membership functions for the quality variable. We show in Figure 2.9 the membership functions for the roughness variable. We also show in Figure 2.10 the use of the "rule viewer" of MATLAB to calculate specific values. Finally, in Figure 2.11 we show the non-linear surface for the Mamdani model.

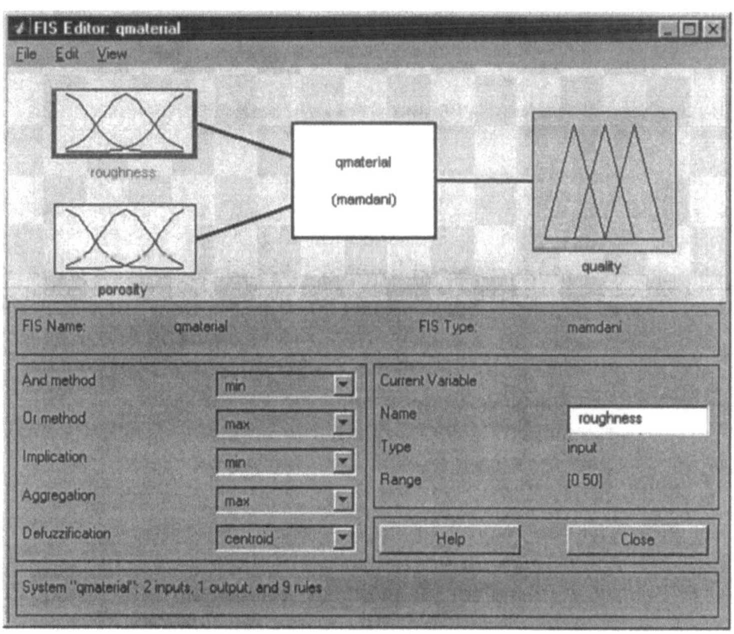

Figure 2.6 Architecture of the fuzzy system for quality evaluation.

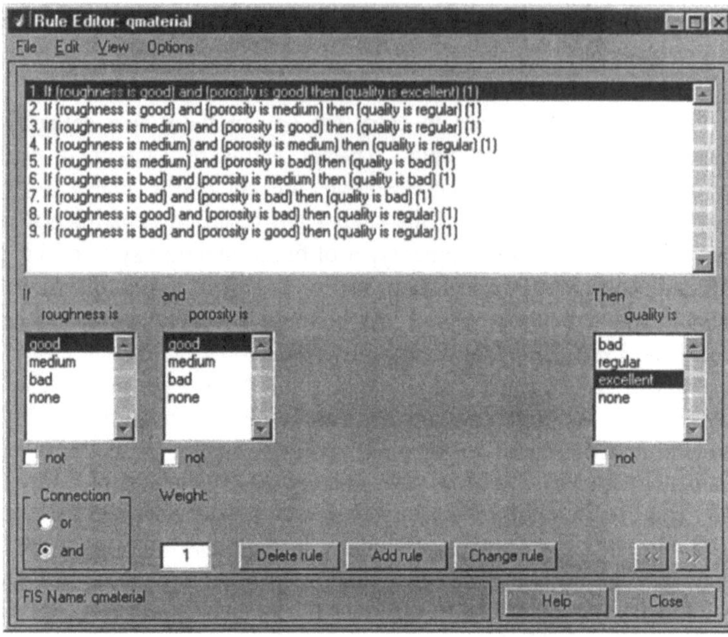

Figure 2.7 Fuzzy rule base for quality evaluation.

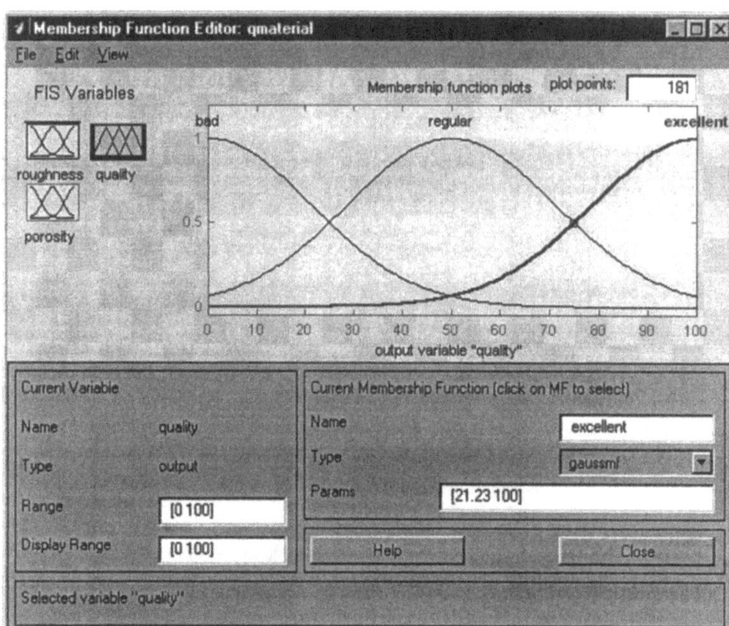

Figure 2.8 Gaussian membership functions for the output linguistic variable.

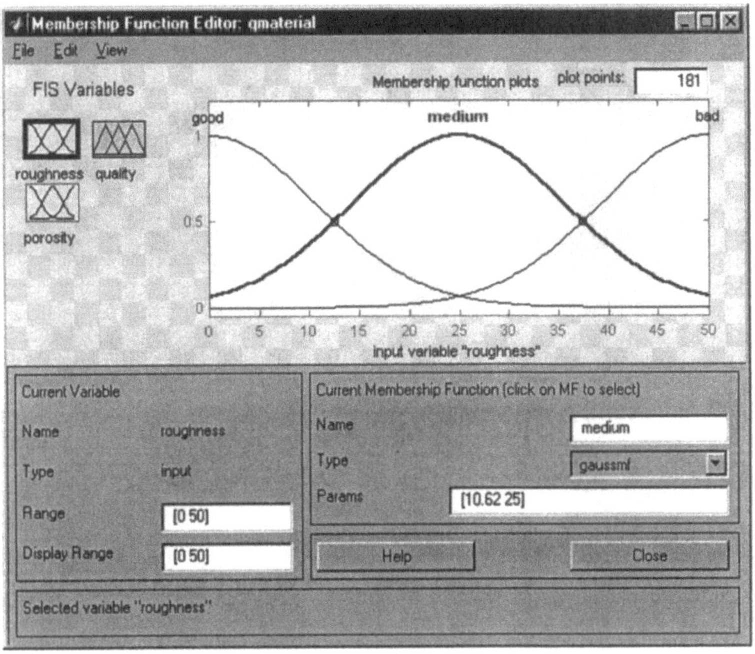

Figure 2.9 Gaussian membership functions for the roughness linguistic variable.

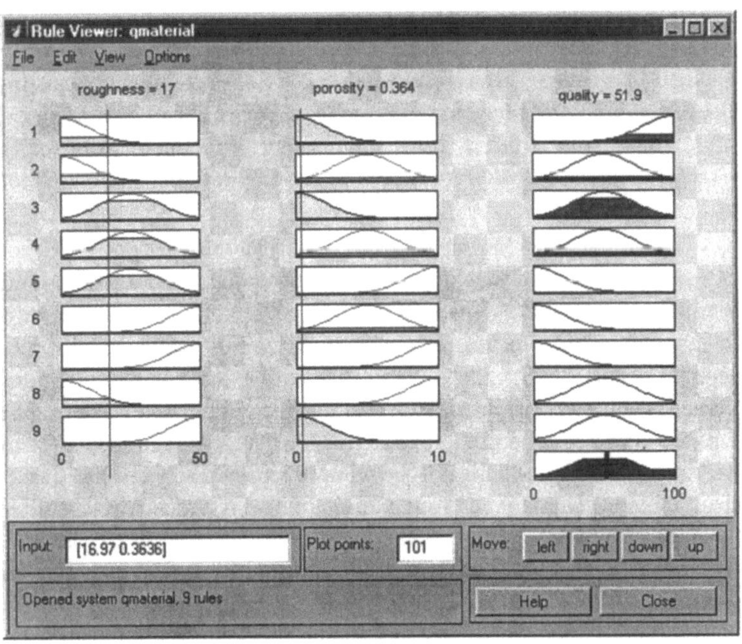

Figure 2.10 Use of the fuzzy rule base with specific values.

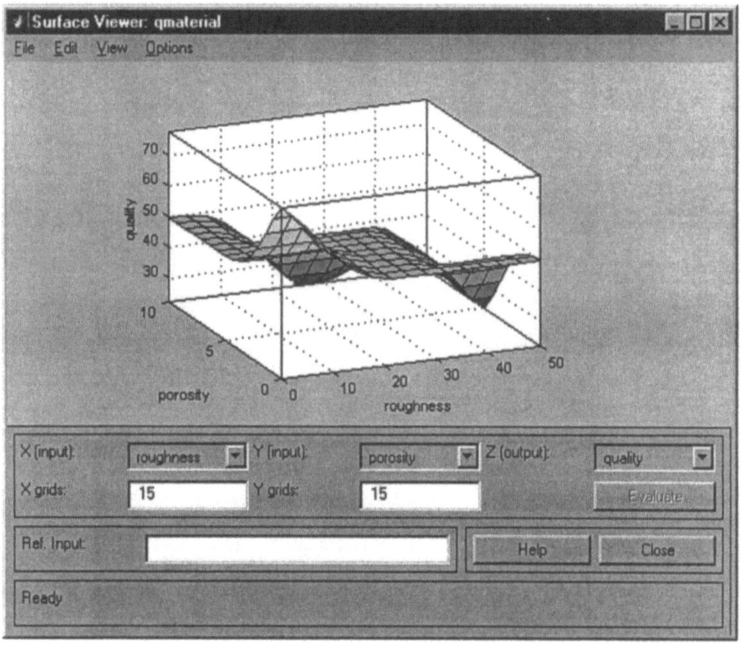

Figure 2.11 Non-linear surface of the Mamdani fuzzy model.

Sugeno Fuzzy Models

The "Sugeno fuzzy model" (also known as the "TSK fuzzy model") was proposed by Takagi, Sugeno and Kang in an effort to develop a systematic approach to generating fuzzy rules from a given input-output data set (Takagi & Sugeno, 1985), (Sugeno & Kang, 1988). A typical fuzzy rule in a Sugeno fuzzy model has the form:

$$\textbf{if } x \text{ is } A \textbf{ and } y \text{ is } B \textbf{ then } z = f(x,y)$$

where A and B are fuzzy sets in the antecedent, while z = f(x,y) is a traditional function in the consequent. Usually f(x,y) is a polynomial in the input variables x and y, but it can be any function as long as it can appropriately describe the output of the model within the fuzzy region specified by the antecedent of the rule. When f(x,y) is a first-order polynomial, the resulting fuzzy inference system is called a "first-order Sugeno fuzzy model". When f is constant, we then have a "zero-order Sugeno fuzzy model", which can be viewed either as a special case of the Mamdani inference system, in which each rule's consequent is specified by a fuzzy singleton, or a special case of the Tsukamoto fuzzy model (to be introduced next), in which each rule's consequent is specified by a MF of a step function center at the constant.

Figure 2.12 shows the fuzzy reasoning procedure for a first-order Sugeno model. Since each rule has a numeric output, the overall output is obtained via "weighted average", thus avoiding the time-consuming process of defuzzification required in a Mamdani model. In practice, the weighted average operator is sometimes replaced with the "weighted sum" operator (that is, $w_1z_1 + w_2z_2$ in Figure 2.12) to reduce computation further specially, in the training of a fuzzy inference system. However, this simplification could lead to the loss of MF linguistic meanings unless the sum of firing strengths (that is, $\sum w_i$) is close to unity.

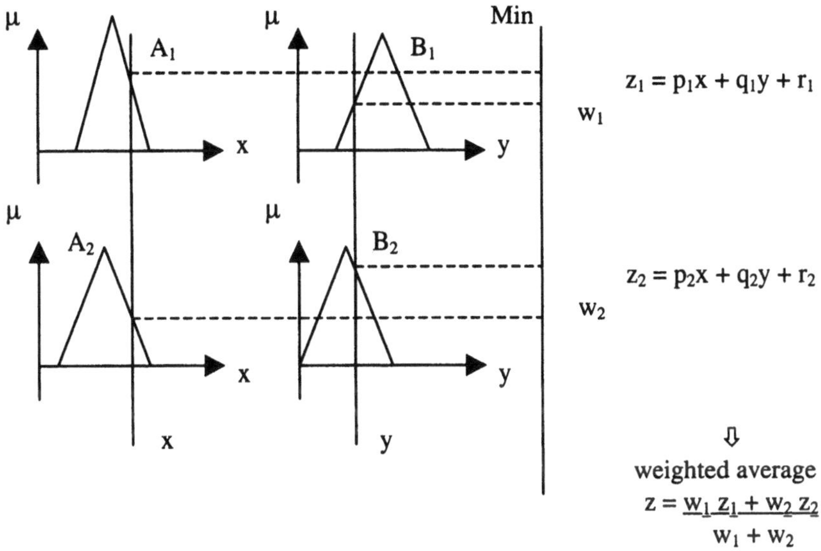

Figure 2.12 The Sugeno fuzzy model.

Unlike the Mamdani fuzzy model, the Sugeno fuzzy model cannot follow the compositional rule of inference strictly in its fuzzy reasoning mechanism. This poses some difficulties when the inputs to a Sugeno fuzzy model are fuzzy. Specifically, we can still employ the matching of fuzzy sets to find the firing strength of each rule. However, the resulting overall output via either weighted average or weighted sum is always crisp; this is counterintuitive since a fuzzy model should be able to propagate the fuzziness from inputs to outputs in an appropriate manner. Without the use of the time-consuming defuzzification procedure, the Sugeno fuzzy model is by far the most popular candidate for sample-data-based modeling, which will be introduced in Chapter 4.

We will give a simple example to illustrate the use of the Sugeno fuzzy inference system. We will consider again the manufacturing example, i.e., determining the quality of a material depending on the roughness and porosity. In Figure 2.13 we show the architecture of the Sugeno model for this example. We show in Figure 2.14 the fuzzy rule base of the Sugeno model. We also show in Figure 2.15 the membership functions for the roughness input variable. In Figure 2.16 we show the non-linear surface of the Sugeno model.

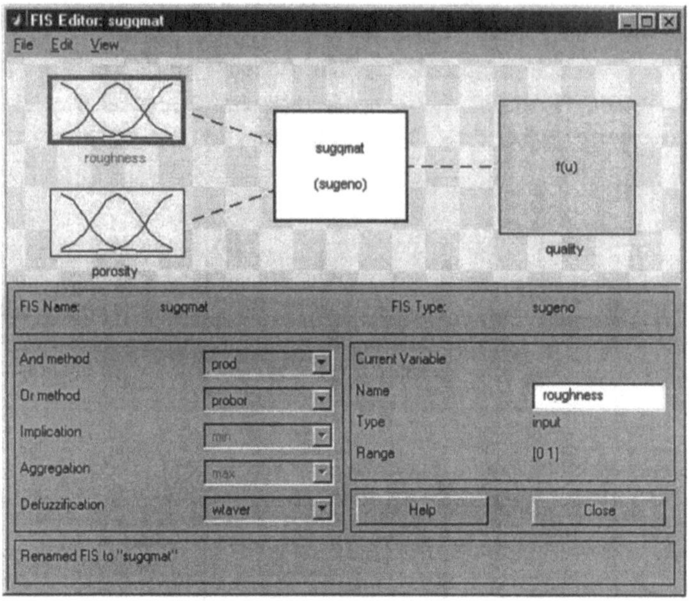

Figure 2.13 Architecture of the Sugeno fuzzy model for quality evaluation.

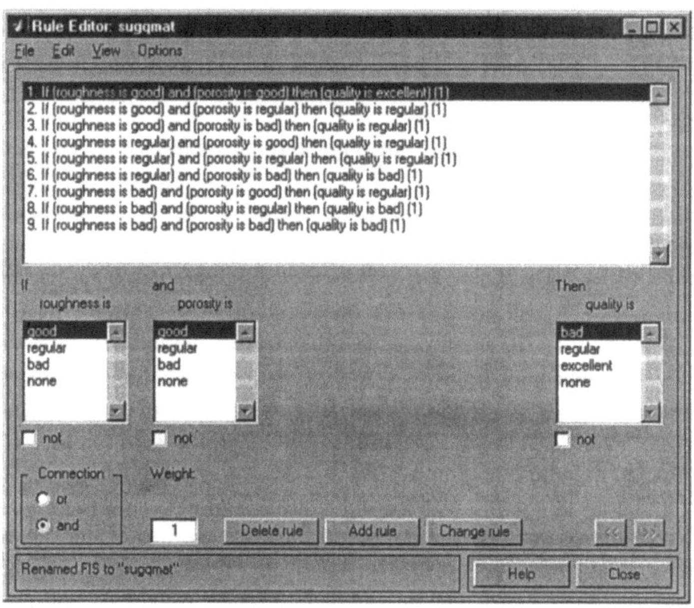

Figure 2.14 Fuzzy rule base for quality evaluation using the "rule editor".

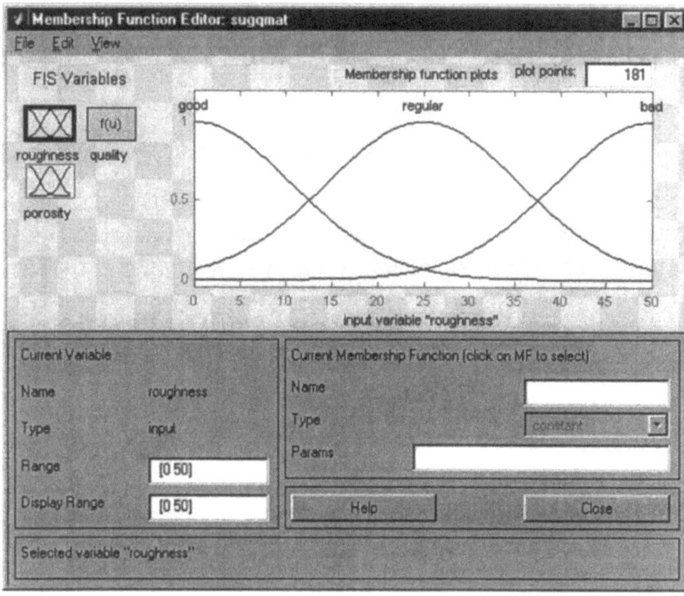

Figure 2.15 Membership functions for the roughness linguistic variable.

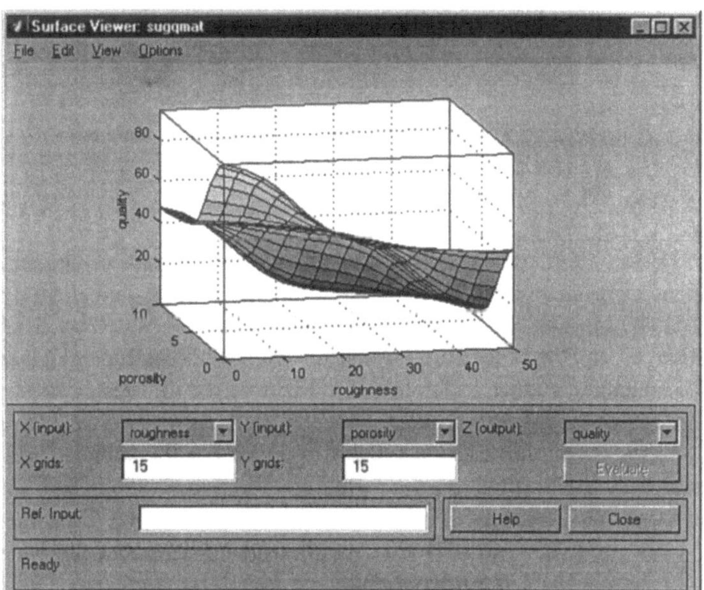

Figure 2.16 Non-linear surface for the Sugeno fuzzy model for quality evaluation.

Finally, we show in Figure 2.17 the use of the "rule viewer" of the Fuzzy Logic Toolbox of MATLAB. The rule viewer is used when we want to evaluate the output of a fuzzy system using specific values for the input variables. In Figure 2.17, for example, we give a roughness of 3.67, and a porosity of 0.0, and obtain as a result a quality of 91.3%, which is excellent.

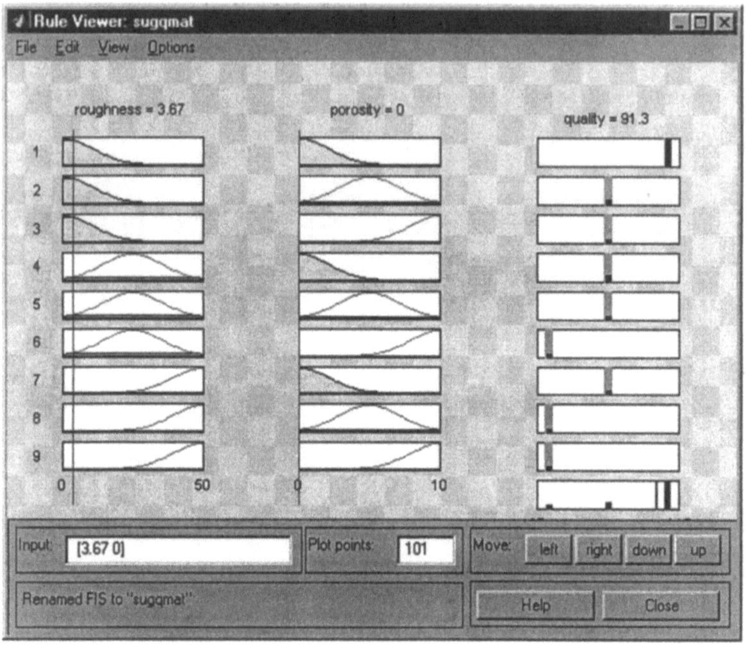

Figure 2.17 Application of the rule viewer of MATLAB with specific values.

<u>Tsukamoto Fuzzy Models</u>

In the "Tsukamoto fuzzy models" (Tsukamoto, 1979), the consequent of each fuzzy if-then rule is represented by a fuzzy set with a monotonical MF, as shown in Figure 2.18. As a result, the inferred output of each rule is defined as a numeric value induced by the rule firing strength. The overall output is taken as the weighted average of each rule's output. Figure 2.18 illustrates the reasoning procedure for a two-input two-rule system.

Since each rule infers a numeric output, the Tsukamoto fuzzy model aggregates each rule's output by the method of weighted average and thus avoids the time-consuming process of defuzzification. However, the Tsukamoto fuzzy model is not used often since it is not as transparent as either the Mamdani or Sugeno fuzzy models. Since the reasoning method of the Tsukamoto fuzzy model does not follow strictly the compositional rule of inference, the output is always crisp even when the inputs are fuzzy.

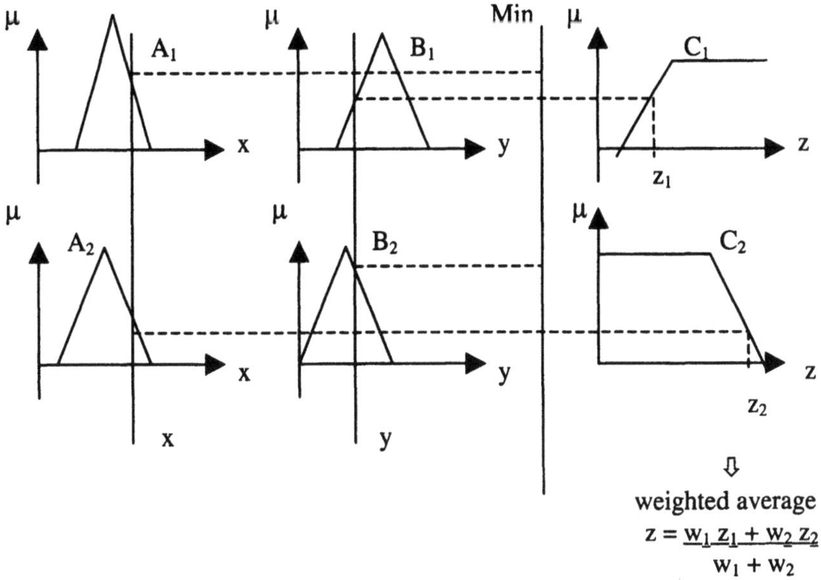

Figure 2.18 The Tsukamoto fuzzy model.

There are certain common issues concerning all the three fuzzy inference systems introduced previously, such as how to partition an input space and how to construct a fuzzy inference system for a particular application. We will examine these issues in more detail in the following lines.

<u>Input Space Partitioning</u>

Now it should be clear that thc main idea of fuzzy inference systems resembles that of "divide and conquer" - the antecedent of a fuzzy rule defines a local fuzzy region, while the consequent describes the behavior within the region via various constituents. The consequent constituent can be a consequent MF (Mamdani and Tsukamoto fuzzy models), a constant value (zero-order Sugeno model), a linear equation (first-order Sugeno model) or a non-linear equation (higher order Sugeno models). Different consequent constituents result in different fuzzy inference systems, but their antecedents are always the same. Therefore, the following discussion of methods of partitioning input spaces to form the antecedents of fuzzy rules is applicable to all three types of fuzzy inference systems.

- Grid partition: This partition method is often chosen in designing a fuzzy controller, which usually involves only several state variables as the inputs to the controller. This partition strategy needs only a small number of MFs for each input. However, it encounters problems when we have a moderately large number of inputs. For instance, a fuzzy model with 12 inputs and 2 MFs on each input would result in $2^{12} = 4096$ fuzzy if-then rules, which is prohibitively large. This problem, usually referred to as the "curse of dimensionality", can be alleviated by other partition strategies.

- Tree partition: In this method each region can be uniquely specified along a corresponding decision tree. The tree partition relieves the problem of an exponential increase in the number of rules. However, more MFs for each input are needed to define these fuzzy regions, and these MFs do not usually bear clear linguistic meanings. In other words, orthogonality holds roughly in $X \times Y$, but not in either X or Y alone.
- Scatter partition: By covering a subset of the whole input space that characterizes a region of possible occurrence of the input vectors, the scatter partition can also limit the number of rules to a reasonable amount. However, the scatter partition is usually dictated by desired input-output data pairs and thus, in general, orthogonality does not hold in X, Y or $X \times Y$. This makes it hard to estimate the overall mapping directly from the consequent of each rule's output.

2.4 Fuzzy Modelling

In general, we design a fuzzy inference system based on the past known behavior of a target system. The fuzzy system is then expected to be able to reproduce the behavior of the target system. For example, if the target system is a human operator in charge of a electrochemical reaction process, then the fuzzy inference system becomes a fuzzy logic controller that can regulate and control the process.

Let us now consider how we might construct a fuzzy inference system for a specific application. Generally speaking, the standard method for constructing a fuzzy inference system, a process usually called "fuzzy modelling", has the following features:

- The rule structure of a fuzzy inference system makes it easy to incorporate human expertise about the target system directly into the modelling process. Namely, fuzzy modelling takes advantage of "domain knowledge" that might not be easily or directly employed in other modelling approaches.
- When the input-output data of a target system is available, conventional system identification techniques can be used for fuzzy modelling. In other words, the use of "numerical data" also plays an important role in "fuzzy modelling", just as in other mathematical modelling methods.

Conceptually, fuzzy modelling can be pursued in two stages, which are not totally disjoint. The first stage is the identification of the "surface structure", which includes the following tasks:

1. Select relevant input and output variables.

2. Choose a specific type of fuzzy inference system.
3. Determine the number of linguistic terms associated with each input and output variables.
4. Design a collection of fuzzy if-then rules.

Note that to accomplish the preceding tasks, we rely on our own knowledge (common sense, simple physical laws, an so on) of the target system, information provided by human experts who are familiar with the target system, or simply trial and error.

After the first stage of fuzzy modelling, we obtain a rule base that can more or less describe the behavior of the target system by means of linguistic terms. The meaning of these linguistic terms is determined in the second stage, the identification of "deep structure", which determines the MFs of each linguistic term (and the coefficients of each rule's output in the case that a Sugeno model is used). Specifically, the identification of deep structure includes the following tasks:

1. Choose an appropriate family of parameterized MFs.
2. Interview human experts familiar with the target systems to determine the parameters of the MFs used in the rule base.
3. Refine the parameters of the MFs using regression and optimization techniques.

Task 1 and 2 assume the availability of human experts, while task 3 assumes the availability of a desired input-output data set. When a fuzzy inference system is used as a controller for a given plant, then the objective in task 3 should be changed to that of searching for parameters that will generate the best performance of the plant.

2.5 Summary

In this chapter, we have presented the main ideas underlying type-1 fuzzy logic and we have only started to point out the many possible applications of this powerful computational theory. We have discussed in some detail fuzzy set theory, fuzzy reasoning and fuzzy inference systems. At the end, we also gave some remarks about fuzzy modelling. In the following chapters, we will show how fuzzy logic techniques (in some cases, in conjunction with other methodologies) can be applied to solve real world complex problems. This chapter will serve as a basis for the new hybrid intelligent methods, for modelling and simulation, that will be described later in this book. Fuzzy Logic will also play an important role in the new neuro-fuzzy methodology for control that is presented later in this book. Applications of fuzzy logic to achieve intelligent manufacturing will be described in Chapters 9, 10, 11, and 12.

Chapter 3

Type-2 Fuzzy Logic

We describe in this chapter a new area in fuzzy logic, which studies type-2 fuzzy sets and type-2 fuzzy logic systems. Basically, a type-2 fuzzy set is a set in which we also have uncertainty about the membership function. Of course, type-2 fuzzy systems consist of fuzzy if-then rules, which contain type-2 fuzzy sets. We can say that type-2 fuzzy logic is a generalization of conventional fuzzy logic (type-1) in the sense that uncertainty is not only limited to the linguistic variables but also is present in the definition of the membership functions.

Fuzzy Logic Systems are comprised of rules. Quite often, the knowledge that is used to build these rules is uncertain. Such uncertainty leads to rules whose antecedents or consequents are uncertain, which translates into uncertain antecedent or consequent membership functions (Karnik & Mendel 1998). Type-1 fuzzy systems (like the ones seen in the previous chapter), whose membership functions are type-1 fuzzy sets, are unable to directly handle such uncertainties. We describe in this chapter, type-2 fuzzy systems, in which the antecedent or consequent membership functions are type-2 fuzzy sets. Such sets are fuzzy sets whose membership grades themselves are type-1 fuzzy sets; they are very useful in circumstances where it is difficult to determine an exact membership function for a fuzzy set.

The original fuzzy logic, founded by Lotfi Zadeh, has been around for more than 30 years, and yet it is unable to handle uncertainties (Mendel, 2001). That the original fuzzy logic (type-1 fuzzy logic) cannot do this sounds paradoxical because the word "fuzzy" has the connotation of uncertainty. The expanded fuzzy logic (type-2 fuzzy logic) is able to handle uncertainties because it can model and minimize their effects.

In what follows, we shall first introduce the basic concepts of type-2 fuzzy sets, and type-2 fuzzy reasoning. Then we will introduce and compare the different types of fuzzy inference systems that have been employed in various applications. We will also consider briefly type-2 fuzzy logic systems and the comparison to type-1 fuzzy systems. Finally, we will address briefly the features and problems of fuzzy modeling with type-2 fuzzy logic, which is concerned with the construction of fuzzy inference systems for modeling a given target system.

3.1 Type-2 Fuzzy Sets

The concept of a type-2 fuzzy set, was introduced by Zadeh (1975) as an extension of the concept of an ordinary fuzzy set (henceforth called a "type-1 fuzzy set"). A type-2 fuzzy set is characterized by a fuzzy membership function, i.e., the membership grade for each element of this set is a fuzzy set in [0,1], unlike a type-1 set where the membership grade is a crisp number in [0,1]. Such sets can be used in situations where there is uncertainty about the membership grades themselves, e.g., an uncertainty in the shape of the membership function or in some of its parameters. Consider the transition from ordinary sets to fuzzy sets. When we cannot determine the membership of an element in a set as 0 or 1, we use fuzzy sets of type-1. Similarly, when the situation is so fuzzy that we have trouble determining the membership grade even as a crisp number in [0,1], we use fuzzy sets of type-2.

This does not mean that we need to have extremely fuzzy situations to use type-2 fuzzy sets. There are many real-world problems where we cannot determine the exact form of the membership functions, e.g., in time series prediction because of noise in the data. Another way of viewing this is to consider type-1 fuzzy sets as a first order approximation to the uncertainty in the real-world. Then type-2 fuzzy sets can be considered as a second order approximation. Of course, it is possible to consider fuzzy sets of higher types but the complexity of the fuzzy system increases very rapidly. For this reason, we will only consider very briefly type-2 fuzzy sets. Lets consider some simple examples of type-2 fuzzy sets.

<u>Example</u> 3.1 Consider the case of a fuzzy set characterized by a Gaussian membership function with mean m and a standard deviation that can take values in $[\sigma_1,\sigma_2]$, i.e.,

$$\mu(x)=\exp\{-\tfrac{1}{2}[(x-m)/\sigma]^2\};\ \sigma\in[\sigma_1,\sigma_2] \qquad (3.1)$$

Corresponding to each value of σ, we will get a different membership curve (see Figure 3.1). So, the membership grade of any particular x (except x=m) can take any of a number of possible values depending upon the value of σ, i.e., the membership grade is not a crisp number, it is a fuzzy set. Figure 3.1 shows the domain of the fuzzy set associated with x=0.7; however, the membership function associated with this fuzzy set is not shown in the figure.

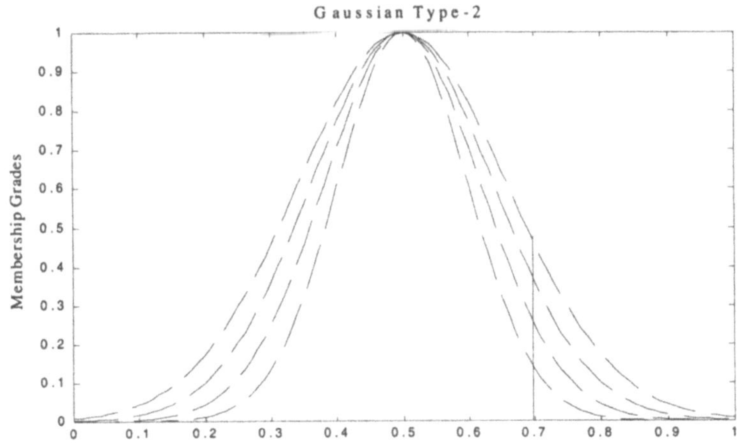

Figure 3.1 A type-2 fuzzy set representing a type-1 fuzzy set with uncertain standard deviation.

Example 3.2 Consider the case of a fuzzy set with a Gaussian membership function having a fixed standard deviation σ, but an uncertain mean, taking values in $[m_1, m_2]$, i.e.,

$$\mu(x)=\exp\{-½[(x-m)/\sigma]^2\};\ m\in[m_1,m_2] \quad (3.2)$$

Again, μ(x) is a fuzzy set. Figure 3.2 shows an example of such a set.

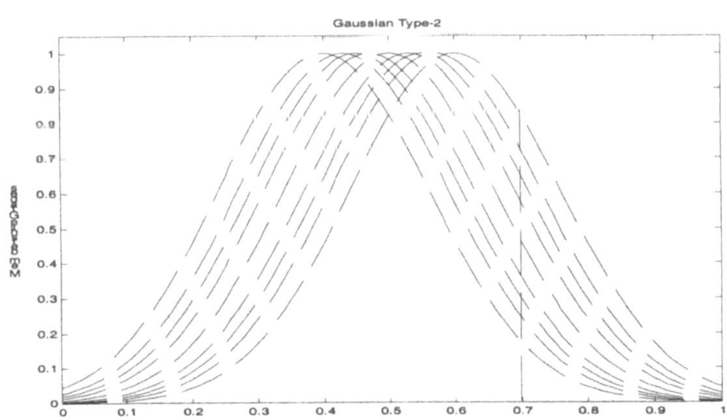

Figure 3.2 A type-2 fuzzy set representing a type-1 fuzzy set with uncertain mean. The mean is uncertain in the interval [0.4, 0.6].

Example 3.3 Consider a type-1 fuzzy set characterized by a Gaussian membership function (mean M and standard deviation σ_x), which gives one crisp membership m(x) for each input x ∈ X, where

$$m(x)=\exp\{-½\,[(x-M)/\sigma_x]^2\} \qquad (3.3)$$

This is shown in Figure 3.3. Now, imagine that this membership of x is a fuzzy set. Let us call the domain elements of this set "primary memberships" of x (denoted by μ_1) and membership grades of these primary memberships "secondary memberships" of x [denoted by $\mu_2(x, \mu_1)$]. So, for a fixed x, we get a type-1 fuzzy set whose domain elements are primary memberships of x and whose corresponding membership grades are secondary memberships of x. If we assume that the secondary memberships follow a Gaussian with mean m(x) and standard deviation σ_m, as in Figure 3.3, we can describe the secondary membership function for each x as

$$\mu_2(x,\mu_1) = e - ½\,[(\mu_1 - m(x))/\,\sigma_m]^2 \qquad (3.4)$$

where $\mu_1 \in [0,1]$ and m is as in equation (3.3).

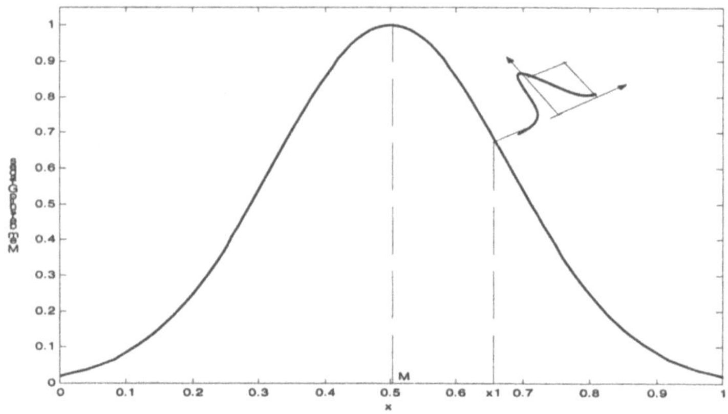

Figure 3.3 A type-2 fuzzy set in which the membership grade of every domain point is a Gaussian type-1 set.

We can formally define these two kinds of type-2 sets as follows.

Definition 3.1 **Gaussian type-2**

A Gaussian type-2 fuzzy set is one in which the membership grade of every domain point is a Gaussian type-1 set contained in [0,1].

Example 3.3 shows an example of a Gaussian type-2 fuzzy set.

Definition 3.2 **Interval type-2**
An interval type-2 fuzzy set is one in which the membership grade of every domain point is a crisp set whose domain is some interval contained in [0,1].

Example 3.1 shows an example of an interval type-2 fuzzy set.

We will give some useful definitions on type-2 fuzzy sets in the following lines.

Definition 3.3 **Footprint of uncertainty**
Uncertainty in the primary memberships of a type-2 fuzzy set, $\tilde{A}$, consists of a bounded region that we call the "footprint of uncertainty" (FOU). Mathematically, it is the union of all primary membership functions (Mendel 2001).

Definition 3.4 **Upper and lower membership functions**
An "upper membership function" and a "lower membership functions" are two type-1 membership functions that are bounds for the FOU of a type-2 fuzzy set $\tilde{A}$. The upper membership function is associated with the upper bound of FOU($\tilde{A}$). The lower membership function is associated with the lower bound of FOU($\tilde{A}$).

We illustrate the concept of upper and lower membership functions as well as the footprint of uncertainty in the following example.

Example 3.4 *Gaussian primary MF with uncertain standard deviation*
For the Gaussian primary membership function with uncertain standard deviation (Figure 3.1), the upper membership function is

$$\text{upper}(\text{FOU}(\tilde{A})) = N(m, \sigma_2; x) \tag{3.5}$$

And the lower membership function is

$$\text{lower}(\text{FOU}(\tilde{A})) = N(m, \sigma_1; x). \tag{3.6}$$

We will describe the operations and properties of type-2 fuzzy sets in the following section.

3.2 Operations on Type-2 Fuzzy Sets

In this section we describe the set theoretic operations of type-2 fuzzy sets. We are interested in the case of type-2 fuzzy sets, $\tilde{A}_i$ (i = 1,...,r), whose secondary membership functions are type-1 fuzzy sets. To compute the union, intersection, and complement of type-2 fuzzy sets, we need to extend the binary operations of minimum (or product) and maximum, and the unary operation of negation, from crisp numbers to type-1 fuzzy sets, because at each x, $\mu_{\tilde{A}i}$ (x, u) is a function (unlike the type-1 case, where $\mu_{\tilde{A}i}$ (x) is a crisp number). The tool for computing the union, intersection, and complement of type-2 fuzzy sets is Zadeh's extension principle (Zadeh, 1975).

Consider two type-2 fuzzy sets $\tilde{A}_1$ and $\tilde{A}_2$, i.e.,

$$\tilde{A}_1 = \int_x \mu_{\tilde{A}1}(x)/\,x \qquad (3.7)$$

and

$$\tilde{A}_2 = \int_x \mu_{\tilde{A}2}(x)/\,x \qquad (3.8)$$

In this section, we focus our attention on set theoretic operations for such general type-2 fuzzy sets.

Definition 3.5 **Union of type-2 fuzzy sets**
The union of $\tilde{A}_1$ and $\tilde{A}_2$ is another type-2 fuzzy set, just as the union of type-1 fuzzy sets A_1 and A_2 is another type-1 fuzzy set. More formally, we have the following expression

$$\tilde{A}_1 \cup \tilde{A}_2 = \int_{x\in X} \mu_{\tilde{A}1\cup\tilde{A}2}(x)/\,x \qquad (3.9).$$

We can explain Equation (3.9) by the "join" operation (Mendel, 2001). Basically, the join between two secondary membership functions must be performed between every possible pair of primary memberships. If more than one combination of pairs gives the same point, then in the join we keep the one with maximum membership grade.

Definition 3.6 **Intersection of type-2 fuzzy sets**
The intersection of $\tilde{A}_1$ and $\tilde{A}_2$ is another type-2 fuzzy set, just as the intersection of type-1 fuzzy sets A_1 and A_2 is another type-1 fuzzy set. More formally, we have the following expression

$$\tilde{A}_1 \cap \tilde{A}_2 = \int_{x\in X} \mu_{\tilde{A}1\cap\tilde{A}2}(x)/\,x \qquad (3.10).$$

We can explain Equation (3.9) by the "meet" operation (Mendel, 2001). Basically, the meet between two secondary membership functions must be performed between every possible pair of primary memberships. If more than one combination of pairs gives the same point, then in the meet we keep the one with maximum membership grade.

Definition 3.7 **Complement of a type-2 fuzzy set**
The complement of set $\tilde{A}$ is another type-2 fuzzy set, just as the complement of type-1 fuzzy set A is another type-1 fuzzy set. More formally we have

$$\tilde{A}' = \int_x \mu_{\tilde{A}'1}(x)/x \qquad (3.11)$$

where the prime denotes complement in the above equation. In this equation $\mu_{\tilde{A}'1}$ is a secondary membership function, i.e., at each value of x $\mu_{\tilde{A}'1}$ is a function (unlike the type-1 case where, at each value of x, $\mu_{\tilde{A}'1}$ is a point value).

Example 3.5 *Type-2 fuzzy set operations*
In this example we illustrate the union, intersection and complement operations for two type-2 fuzzy sets $\tilde{A}_1$ and $\tilde{A}_2$, and for a particular element x for which the secondary membership functions in these two sets are $\mu_{\tilde{A}1}(x) = 0.5/0.1 + 0.8/0.2$ and $\mu_{\tilde{A}2}(x) = 0.4/0.5 + 0.9/0.9$. Using in the operations the minimum t-norm and the maximum t-conorm, we have the following results:

$$\begin{aligned}\mu_{\tilde{A}1\cup\tilde{A}2}(x) = \mu_{\tilde{A}1}(x) \cup \mu_{\tilde{A}2}(x) &= (0.5/0.1 + 0.8/0.2) \cup (0.4/0.5 + 0.9/0.9)\\ &= (0.5 \wedge 0.4)/(0.1 \vee 0.5) + (0.5 \wedge 0.9)/(0.1 \vee 0.9) +\\ &\quad (0.8 \wedge 0.4)/(0.2 \vee 0.5) + (0.8 \wedge 0.9)/(0.2 \vee 0.9)\\ &= 0.4/0.5 + 0.5/0.9 + 0.4/0.5 + 0.8/0.9\\ &= \max\{0.4, 0.4\}/0.5 + \max\{0.5, 0.8\}/0.9\\ &= 0.4/0.5 + 0.8/0.9\end{aligned}$$

$$\begin{aligned}\mu_{\tilde{A}1\cap\tilde{A}2}(x) = \mu_{\tilde{A}1}(x) \cap \mu_{\tilde{A}2}(x) &= (0.5/0.1 + 0.8/0.2) \cap (0.4/0.5 + 0.9/0.9)\\ &= (0.5 \wedge 0.4)/(0.1 \wedge 0.5) + (0.5 \wedge 0.9)/(0.1 \wedge 0.9) +\\ &\quad (0.8 \wedge 0.4)/(0.2 \wedge 0.5) + (0.8 \wedge 0.9)/(0.2 \wedge 0.9)\\ &= 0.4/0.1 + 0.5/0.1 + 0.4/0.2 + 0.8/0.2\\ &= \max\{0.4, 0.5\}/0.1 + \max\{0.4, 0.8\}/0.2\\ &= 0.5/0.1 + 0.8/0.2\end{aligned}$$

$$\mu_{\tilde{A}'1}(x) = 0.5/(1 - 0.1) + 0.8/(1 - 0.2) = 0.5/0.9 + 0.8/0.8.$$

3.3 Type-2 Fuzzy Systems

The basics of fuzzy logic do not change from type-1 to type-2 fuzzy sets, and in general, will not change for any type-n (Karnik & Mendel 1998). A higher-type number just indicates a higher "degree of fuzziness". Since a higher type changes the nature of the membership functions, the operations that depend on the membership functions change; however, the basic principles of fuzzy logic are independent of the nature of membership functions and hence, do not change. Rules of inference like Generalized Modus Ponens or Generalized Modus Tollens continue to apply.

The structure of the type-2 fuzzy rules is the same as for the type-1 case because the distinction between type-2 and type-1 is associated with the nature of the membership functions. Hence, the only difference is that now some or all the sets involved in the rules are of type-2. In a type-1 fuzzy system, where the output sets are type-1 fuzzy sets, we perform defuzzification in order to get a number, which is in some sense a crisp (type-0) representative of the combined output sets. In the type-2 case, the output sets are type-2; so we have to use extended versions of type-1 defuzzification methods. Since type-1 defuzzification gives a crisp number at the output of the fuzzy system, the extended defuzzification operation in the type-2 case gives a type-1 fuzzy set at the output. Since this operation takes us from the type-2 output sets of the fuzzy system to a type-1 set, we can call this operation "type reduction" and call the type-1 fuzzy set so obtained a "type-reduced set". The type-reduced fuzzy set may then be defuzzified to obtain a single crisp number; however, in many applications, the type-reduced set may be more important than a single crisp number.

Type-2 sets can be used to convey the uncertainties in membership functions of type-1 fuzzy sets, due to the dependence of the membership functions on available linguistic and numerical information. Linguistic information (e.g. rules from experts), in general, does not give any information about the shapes of the membership functions. When membership functions are determined or tuned based on numerical data, the uncertainty in the numerical data, e.g., noise, translates into uncertainty in the membership functions. In all such cases, any available information about the linguistic/numerical uncertainty can be incorporated in the type-2 framework. However, even with all of the advantages that fuzzy type-2 systems have, the literature on the applications of type-2 sets is scarce. Some examples are Yager (1980) for decision making, and Wagenknecht & Hartmann (1988) for solving fuzzy relational equations. We think that more applications of type-2 fuzzy systems will come in the near future as the area matures and the theoretical results become more understandable for the general public in the fuzzy arena.

3.3.1 Singleton Type-2 Fuzzy Logic Systems

This section discusses the structure of a singleton type-2 Fuzzy Logic Systems (FLS), which is a system that accounts for uncertainties about the antecedents or consequents in rules, but does not explicitly account for input measurement uncertainties. More complicated (but, more versatile) non-singleton type-2 FLSs, which account for both types of uncertainties, are discussed later.

The basics of fuzzy logic do not change from type-1 to type-2 fuzzy sets, and in general will not change for type-n. A higher type number just indicates a higher degree of fuzziness. Since a higher type changes the nature of the membership functions, the operations that depend on the membership functions change, however, the basic principles of fuzzy logic are independent of the nature of membership functions and hence do not change. Rules of inference, like Generalized Modus Ponens, continue to apply.

A general type-2 FLS is shown in Figure 3.4. As discussed before a type-2 FLS is very similar to type-1 FLS, the major structural difference being that the defuzzifier block of a type-1 FLS is replaced by the output processing block in type-2 FLS. That block consists of type-reduction followed by defuzzification.

During our development of a type-2 FLS, we assume that all the antecedent and consequent sets in rules are type-2, however, this need not necessarily be the case in practice. All results remain valid as long as long as just one set is type-2. This means that a FLS is type-2 as long as any one of its antecedent or consequent sets is type-2.

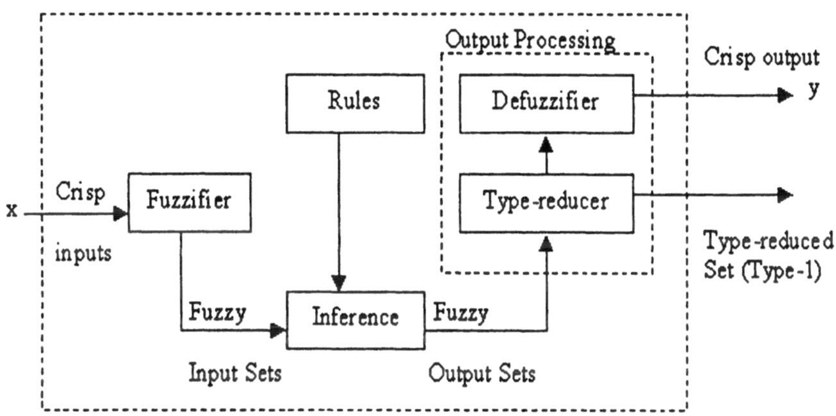

Figure 3.4 Type-2 Fuzzy Logic System.

In the type-1 case, we generally have fuzzy if-then rules of the form

$$R^l: \text{IF } x_1 \text{ is } A^l_1 \text{ and } \ldots\ x_p \text{ is } A^l_p, \text{ THEN } y \text{ is } Y^l \qquad l = 1,\ldots, M \quad (3.12)$$

As mentioned earlier, the distinction between type-1 and type-2 is associated with the nature of the membership functions, which is not important when forming the rules. The structure of the rules remains exactly the same in the type-2 case, but now some or all of the sets involved are type-2.

Consider a type-2 FLS having r inputs $x_1 \in X_1, \ldots, x_r \in X_r$ and one output $y \in Y$. As in the type-1 case, we can assume that there are M rules; but, in the type-2 case the *l*th rule has the form

$$R^l: \text{IF } x_1 \text{ is } \tilde{A}^l_1 \text{ and } \ldots\ x_p \text{ is } \tilde{A}^l_p, \text{ THEN } y \text{ is } \hat{Y}^l \qquad l = 1,\ldots, M \quad (3.12)$$

This rule represents a type-2 fuzzy relation between the input space $X_1 \times \ldots \times X_r$, and the output space, Y, of the type-2 fuzzy system.

In a type-1 FLS the inference engine combines rules and gives a mapping from input type-1 fuzzy sets to output type-1 fuzzy sets. Multiple antecedents in rules are combined by the t-norm. The membership grades in the input sets are combined with those in the output sets using composition. Multiple rules may be combined using the t-conorm or during defuzzification by weighted summation. In the type-2 case the inference process is very similar. The inference engine combines rules and gives a mapping from input type-2 fuzzy sets to output type-2 fuzzy sets. To do this one needs to compute unions and intersections of type-2 fuzzy sets, as well as compositions of type-2 relations.

In the type-2 fuzzy system (Figure 3.4), as in the type-1 fuzzy system, crisp inputs are first fuzzified into fuzzy input sets that then activate the inference block, which in the present case is associated with type-2 fuzzy sets. In this section, we describe singleton fuzzification and the effect of such fuzzification on the inference engine. The "fuzzifier" maps a crisp point $\mathbf{x} = (x_1,\ldots, x_r)^T \in X_1 \times X_2 \ldots \times X_r \equiv \mathbf{X}$ into a type-2 fuzzy set $\tilde{A}_x$ in $\mathbf{X}$.

The type-2 output of the inference engine shown in Figure 3.4 must be processed next by the output processor, the first operation of which is type-reduction. Type-reduction methods include (Mendel, 2001): centroid, center-of-sums, height, modified height, and center-of-sets. Lets assume that we perform centroid type-reduction. Then each element of the type-reduced set is the centroid of some embedded type-1 set for the output type-2 set of the FLS. Each of these embedded sets can be thought of as an output set of an associated type-1 FLS, and, correspondingly, the type-2 FLS can be viewed of as a collection of many different type-1 FLSs. Each type-1 FLS is embedded in the type-2 FLS; hence, the type-reduced set is a collection of the outputs of all of the embedded type-1 FLSs (see Figure 3.5). The type-reduced set lets us represent the output of the type-2 FLS as a fuzzy set rather than as a crisp number, which is something that cannot be done with a type-1 fuzzy system.

Referring to Figure 3.5, when the antecedent and consequent membership functions of the type-2 FLS have continuous domains, the number of embedded sets is uncountable. Figure 3.5 shows a situation in which we have assumed that the membership functions have discrete (or discretized) domains. The memberships in the type-reduced set, $\mu_Y(y_i)$, represent the level of uncertainty associated with each embedded type-1 FLS. A crisp output can be obtained by aggregating the outputs of all embedded type-1 FLSs by, e.g., finding the centroid of the type-reduced set.

If all of the type-2 uncertainties were to disappear, the secondary membership functions for all antecedents and consequents would each collapse to a single point, which shows that the type-2 FLS collapses to a type-1 FLS.

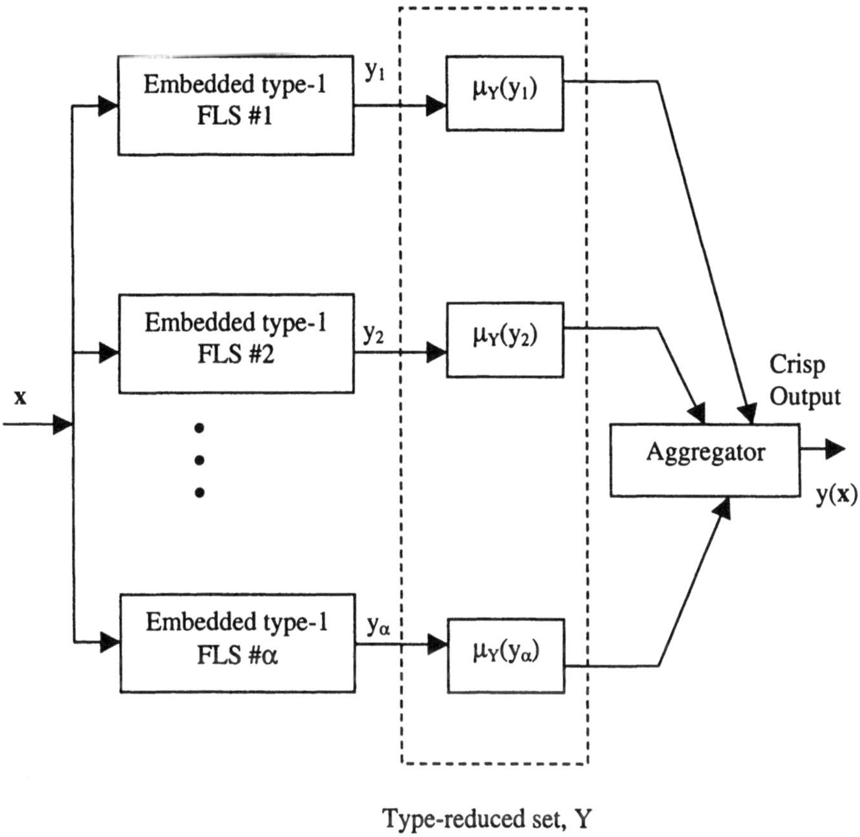

Figure 3.5 A type-2 FLS viewed as a collection of embedded type-1 FLSs.

If we think of a type-2 FLS as a "perturbation" of a type-1 FLS, due to uncertainties in their membership functions, then the type-reduced set of the type-2 FLS can be thought of as representing the uncertainty in the crisp output due to the perturbation. Some measure of the spread of the type-reduced set may then be taken to indicate the possible variation in the crisp output due to the perturbation. This is analogous to using confidence intervals in a stochastic-uncertainty situation.

We defuzzify the type-reduced set to get a crisp output from the type-2 FLS. The most natural way to do this seems to be finding the centroid of the type-reduced set. Finding the centroid is equivalent to finding the weighted average of the outputs of all the type-1 FLSs that are embedded in the type-2 FLS, where the weights correspond to the memberships in the type-reduced set (see Figure 3.5). If the type-reduced set Y for an input **x** is discretized or is discrete and consists of α points, then the expression for its centroid is

$$y(\mathbf{x}) = [\Sigma^{\alpha}_{k=1}\, y_k\, \mu_Y(y_k)\,]/\,[\Sigma^{\alpha}_{k=1}\, \mu_Y(y_k)\,] \tag{3.13}$$

If α is large then data storage may be a problem for the computation of Equation (3.13). This equation can, however, be evaluated using parallel processing, in this case data storage will not be problem. Currently, however, most researchers still depend on software for simulations and cannot make use of parallel processing. We can, however, use a recursive method to vastly reduce the memory required for storing the data that are needed to compute the defuzzification output. From Equation (3.13), we can calculate

$$A(i) = A(i-1) + y_i\, \mu_Y(y_i) \qquad A(0) = 0 \tag{3.14}$$

and

$$B(i) = B(i-1) + y_i\, \mu_Y(y_i) \qquad B(0) = 0 \tag{3.15}$$

for $i = 1, \ldots, \alpha$. With these formulas we just need to store A and B during each iteration.

From our previous discussions about the five elements that comprise the Figure 3.4 type-2 FLS, we see that there are many possibilities to choose from, even more than for a type-1 FLS. To begin, we must decide on the kind of defuzzification (singleton or non-singleton). We must also choose a FOU for each type-2 membership function, decide on the functional forms for both the primary and secondary membership functions, and choose the parameters of the membership functions (fixed a-priori or tuned during a training procedure). Then we need to choose the composition (max-min, max-product), implication (minimum, product), type-reduction method (centroid, center-of-sums, height, modified height, center-of-sets), and defuzzifier. Clearly, there is an even greater richness among type-2 FLSs than there is among type-1 FLSs. In other words, there are more design degrees of freedom associated with a type-2 FLS than with a type-1 FLS; hence, a type-2 FLS has the potential to outperform a type-1 FLS because of the extra degrees of freedom.

3.3.2 Non-Singleton Fuzzy Logic Systems

A non-singleton FLS is one whose inputs are modeled as fuzzy numbers. A type-2 FLS whose inputs are modeled as type-1 fuzzy numbers is referred to as "type-1 non-singleton type-2 FLS". This kind of a fuzzy system not only accounts for uncertainties about either the antecedents or consequents in rules, but also accounts for input measurement uncertainties.

A type-1 non-singleton type-2 FLS is described by the same diagram as in singleton type-2 FLS, see Figure 3.4. The rules of a type-1 non-singleton type-2 FLS are the same as those for the singleton type-2 FLS. What are different is the fuzzifier, which treats the inputs as type-1 fuzzy sets, and the effect of this on the inference block. The output of the inference block will again be a type-2 fuzzy set; so, the type-reducers and defuzzifier that we described for a singleton type-2 FLS apply as well to a type-1 non-singleton type-2 FLS.

We can also have a situation in which the input are modeled as type-2 fuzzy numbers. This situation can occur, e.g., in time series forecasting when the additive measurement noise is non-stationary. A type-2 FLS whose inputs are modeled as type-2 fuzzy numbers is referred to as "type-2 non-singleton type-2 FLS".

A type-2 non-singleton type-2 FLS is described by the same diagram as in singleton type-2 FLS, see Figure 3.4. The rules of a type-2 non-singleton type-2 FLS are the same as those for a type-1 non-singleton type-2 FLS, which are the same as those for a singleton type-2 FLS. What is different is the fuzzifier, which treats the inputs as type-2 fuzzy sets, and the effect of this on the inference block. The output of the inference block will again be a type-2 fuzzy set; so, the type-reducers and defuzzifier that we described for a type-1 non-singleton type-2 FLS apply as well to a type-2 non-singleton type-2 FLS.

3.3.3 Sugeno Type-2 Fuzzy Systems

All of our previous FLSs were of the Mamdani type, even though we did not refer to them as such. In this section, we will need to distinguish between the two kinds of FLSs, we refer to our previous FLSs as "Mamdani" FLSs. Both kinds of FLS are characterized by if-then rules and have the same antecedent structures. They differ in the structures of their consequents. The consequent of a Mamdani rule is a fuzzy set, while the consequent of a Sugeno rule is a function.

A type-1 Sugeno FLS was proposed by Takagi and Sugeno (1985), and Sugeno and Kang (1988), in an effort to develop a systematic approach to generating fuzzy rules from a given input-output data set. We will consider in this section the extension of first-order type-1 Sugeno FLS to its type-2 counterpart, with emphasis on interval sets.

Consider a type-2 Sugeno FLS having r inputs $x_1 \in X_1, \ldots, x_r \in X_r$ and one output $y \in Y$. A type-2 Sugeno FLS is also described by fuzzy if-then rules that represent input-output relations of a system. In a general first-order type-2 Sugeno model with a rule base of M rules, each having r antecedents, the *i*th rule can be expressed as

$$R^l: \text{IF } x_1 \text{ is } \tilde{A}^l_1 \text{ and } \ldots x_p \text{ is } \tilde{A}^l_p, \quad \text{THEN } Y^i = C^i_0 + C^i_1 x_1 + \ldots + C^i_r x_r \tag{3.16}$$

where $i = 1,\ldots, M$; C^i_j $(j = 1,\ldots, r)$ are consequent type-1 fuzzy sets; Y^i, the output of the *i*th rule, is also a type-1 fuzzy set (because it is a linear combination of type-1 fuzzy sets); and $\tilde{A}^i_k$ $(k = 1, \ldots, r)$ are type-2 antecedent fuzzy sets. These rules let us simultaneously account for uncertainty about antecedent membership functions and consequent parameter values. For a type-2 Sugeno FLS there is no need for type-reduction, just as there is no need for defuzzification in a type-1 Sugeno FLS.

3.4 Summary

In this chapter, we have presented the main ideas underlying type-2 fuzzy logic and we have only started to point out the many possible applications of this powerful computational theory. We have discussed in some detail type-2 fuzzy set theory, fuzzy reasoning and fuzzy inference systems. At the end, we also gave some remarks about type-2 fuzzy modeling with the Mamdani and Sugeno approaches. In the following chapters, we will show how type-2 fuzzy logic (in some cases, in conjunction with other methodologies) can be applied to solve real world complex problems. This chapter will serve as a basis for the new hybrid intelligent methods, for modeling, simulation, and control that will be described later this book.

Chapter 4

Supervised Learning Neural Networks

Application of fuzzy inference systems to automatic control was first reported in Mamdani's paper (Mamdani & Assilian, 1975), where a "fuzzy logic controller" (FLC) was used to emulate a human operator's control of a steam engine and boiler combination. Since then, "fuzzy logic control" has been recognized as the most significant and fruitful application for fuzzy logic (Kosko, 1992). In the past few years, advances in microprocessors and hardware technologies have created an even more diversified application domain for fuzzy logic controllers, which ranges from consumer electronics to the automobile industry. However, without adaptive capability, the performance of fuzzy systems relies exclusively on two factors: the availability of human experts, and the knowledge acquisition techniques to convert human expertise into appropriate fuzzy rules. These two factors substantially restrict the application domain of fuzzy systems.

On the other hand, investigation into using neural networks in industrial systems did not receive much attention until the "backpropagation" learning rule was formulated by Rumelhart and others (Rumelhart, Hinton & Williams, 1986). Since then, research on neural networks has evolved quickly and a number of neural network architectures and design methods have been proposed in the literature (Werbos, 1991).

In this chapter, we present the basic concepts, notation, and basic learning algorithms for supervised neural networks that will be needed in the following chapters of this book. The chapter is organized as follows: backpropagation for feedforward networks, radial basis networks, adaptive neuro-fuzzy inference systems (ANFIS) and applications. First, we give a brief review of the basic concepts of neural networks and the basic backpropagation learning algorithm. Second, we give a brief description of the momentum and adaptive momentum learning algorithms. Third, we give a brief review of the radial basis neural networks. Finally, we end the chapter with a description of the adaptive neuro-fuzzy inference system (ANFIS) methodology. We consider this material necessary to understand the new methods for intelligent manufacturing that will be presented in the final chapters of this book.

4.1 Backpropagation for Feedforward Networks

This section describes the architectures and learning algorithms for adaptive networks, a unifying framework that subsumes almost all kinds of neural network paradigms with supervised learning capabilities. An adaptive network, as the name indicates, is a network structure consisting of a number of nodes connected through directional links. Each node represents a process unit, and the links between nodes specify the causal relationship between the connected nodes. The learning rule specifies how the parameters (of the nodes) should be updated to minimize a prescribed error measure.

The basic learning rule of the adaptive network is the well-known steepest descent method, in which the gradient vector is derived by successive invocations of the chain rule. This method for systematic calculation of the gradient vector was proposed independently several times, by Bryson and Ho (1969), Werbos (1974), and Parker (1982). However, because research on artificial neural networks was still in its infancy at those times, these researchers' early work failed to receive the attention it deserved. In 1986, Rumelhart et al. used the same procedure to find the gradient in a multilayer neural network. Their procedure was called "backpropagation learning rule", a name which is now widely known because the work of Rumelhart inspired enormous interest in research on neural networks. In this section, we introduce Werbos's original backpropagation method for finding gradient vectors and also present improved versions of this method.

4.1.1 The Backpropagation Learning Algorithm

The procedure for finding a gradient vector in a network structure is generally referred to as "backpropagation" because the gradient vector is calculated in the direction opposite to the flow of the ouput of each node. Once the gradient is obtained, a number of derivative-based optimization and regression techniques are available for updating the parameters. In particular, if we use the gradient vector in a simple steepest descent method, the resulting learning paradigm is referred to as "backpropagation learning rule". Suppose that a given feedforward adaptive network has L layers and layer l ($l = 0, 1, \ldots, L$) has N(l) nodes. Then the output and function of node i [$i = 1, \ldots, N(l)$] in layer l can be represented as $x_{l,i}$ and $f_{l,i}$, respectively, as shown in Figure 4.1. Since the output of a node depends on the incoming signals and the parameter set of the node, we have the following general expression for the node function $f_{l,i}$:

$$X_{l,i} = f_{l,i}\,(x_{l-1,1} , \ldots , x_{l-1,N(l-1)}, \alpha, \beta, \gamma , \ldots) \tag{4.1}$$

where α, β, γ , etc. are the parameters of this node.

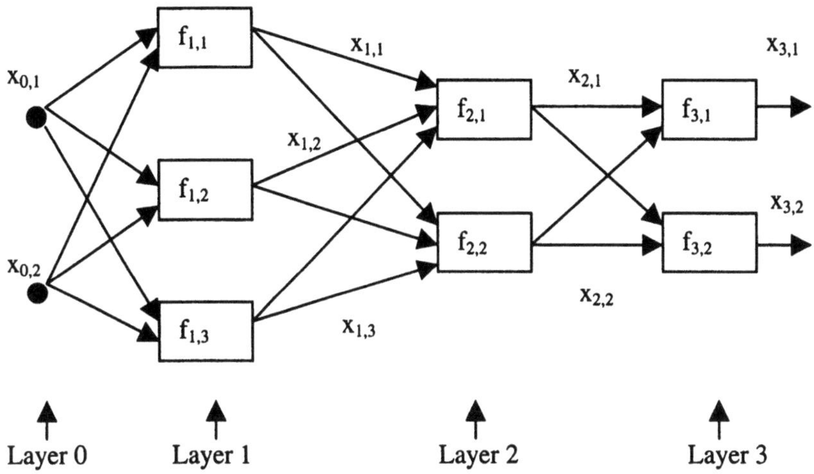

Figure 4.1 Feedforward adaptive network.

Assuming that the given training data set has P entries, we can define an error measure for the pth ($1 \leq p \leq P$) entry of the training data as the sum of the squared errors:

$$E_p = \sum_{k=1}^{N(L)} (d_k - x_{L,k})^2 \tag{4.2}$$

where d_k is the kth component of the pth desired output vector and $x_{L,k}$ is the kth component of the actual output vector produced by presenting the pth input vector to the network. Obviously, when E_p is equal to zero, the network is able to reproduce exactly the desired output vector in the pth training data pair. Thus our task here is to minimize an overall error measure, which is defined as $E = \sum E_p$.

We can also define the "error signal" $\varepsilon_{l,i}$ as the derivative of the error measure E_p with respect to the output of the node i in layer l, taking both direct and indirect paths into consideration. Mathematically,

$$\varepsilon_{l,i} = \frac{\partial^+ E_p}{\partial x_{l,i}} \tag{4.3}$$

this expression was called the "ordered derivative" by Werbos (1974). The difference between the ordered derivative and the ordinary partial derivative lies in the way we view the function to be differentiated. For an internal node output $x_{l,i}$, the partial derivative $\partial^+ E_p / \partial x_{l,i}$ is equal to zero, since E_p does not depend on $x_{l,i}$ directly. However, it is obvious that E_p does depend on $x_{l,i}$ indirectly, since a change in $x_{l,i}$ will propagate through indirect paths to the output layer and thus produce a corresponding change in the value of E_p.

The error signal for the ith output node (at layer L) can be calculated directly:

$$\varepsilon_{L,i} = \frac{\partial^+ E_p}{\partial x_{L,i}} = \frac{\partial E_p}{\partial x_{L,i}} \tag{4.4}$$

This is equal to $\varepsilon_{L,i} = -2(d_i - x_{L,i})$ if E_p is defined as in Equation (4.2). For the internal node at the ith position of layer l, the error signal can be derived by the chain rule of differential calculus:

$$\underbrace{\varepsilon_{l,i} = \frac{\partial^+ E_p}{\partial x_{l,i}}}_{\text{error signal at layer l}} = \sum_{m=1}^{N(l+1)} \underbrace{\frac{\partial^+ E_p}{\partial x_{l+1,m}}}_{\text{error signal at layer l+1}} \frac{\partial f_{l+1,m}}{\partial x_{l,i}} = \sum_{m=1}^{N(l+1)} \varepsilon_{l+1,m} \frac{\partial f_{l+1,m}}{\partial x_{l,i}} \tag{4.5}$$

where $0 \le l \le L-1$. That is, the error signal of an internal node at layer l can be expressed as a linear combination of the error signal of the nodes at layer l+1. Therefore, for any l and i, we can find $\varepsilon_{l,i}$ by first applying Equation (4.4) once to get error signals at the output layer, and then applying Equation (4.5) iteratively until we reach the desired layer l. The underlying procedure is called backpropagation since the error signals are obtained sequentially from the output layer back to the input layer.

The gradient vector is defined as the derivative of the error measure with respect to each parameter, so we have to apply the chain rule again to find the gradient vector. If α is a parameter of the ith node at layer l, we have

$$\frac{\partial^+ E_p}{\partial \alpha} = \frac{\partial^+ E_p}{\partial x_{l,i}} \frac{\partial f_{l,i}}{\partial \alpha} = \varepsilon_{l,i} \frac{\partial f_{l,i}}{\partial \alpha} \tag{4.6}$$

The derivative of the overall error measure E with respect to α is

$$\frac{\partial^+ E}{\partial \alpha} = \sum_{p=1}^{P} \frac{\partial^+ E_p}{\partial \alpha} \tag{4.7}$$

Accordingly, for simple steepest descent (for minimization), the update formula for the generic parameter α is

$$\Delta\alpha = -\eta \frac{\partial^+ E}{\partial \alpha} \tag{4.8}$$

in which η is the "learning rate", which can be further expressed as

$$\eta = \frac{k}{\sqrt{\sum_\alpha (\partial E / \partial \alpha)^2}} \tag{4.9}$$

where k is the "step size", the length of each transition along the gradient direction in the parameter space.

There are two types of learning paradigms that are available to suit the needs for various applications. In "off-line learning" (or "batch learning"), the update formula for parameter α is based on Equation (4.7) and the update action takes place only after the whole training data set has been presented-that is, only after each "epoch" or "sweep". On the other hand, in "on-line learning" (or "pattern-by-pattern learning"), the parameters are updated immediately after each input-output pair has been presented, and the update formula is based on Equation (4.6). In practice, it is possible to combine these two learning modes and update the parameter after k training data entries have been presented, where k is between 1 and P and it is sometimes referred to as the "epoch size".

4.1.2 Backpropagation Multilayer Perceptrons

Artificial neural networks, or simply "neural networks" (NNs), have been studied for more than three decades since Rosenblatt first applied single-layer "perceptrons" to pattern classification learning (Rosenblatt, 1962). However, because Minsky and Papert pointed out that single-layer systems were limited and expressed pessimism over multilayer systems, interest in NNs dwindled in the 1970s (Minsky & Papert, 1969). The recent resurgence of interest in the field of NNs has been inspired by new developments in NN learning algorithms (Fahlman & Lebiere, 1990), analog VLSI circuits, and parallel processing techniques (Lippmann, 1987).

Quite a few NN models have been proposed and investigated in recent years. These NN models can be classified according to various criteria, such as their learning methods (supervised versus unsupervised), architectures (feedforward versus recurrent), output types (binary versus continuous), and so on. In this section, we confine our scope to modelling problems with desired input-output data sets, so the resulting networks must have adjustable parameters that are updated by a supervised learning rule. Such networks are often referred to as "supervised learning" or "mapping networks", since we are interested in shaping the input-output mappings of the network according to a given training data set.

A backpropagation "multilayer perceptron" (MLP) is an adaptive network whose nodes (or neurons) perform the same function on incoming signals; this node function is usually a composite of the weighted sum and a differentiable non-linear activation function, also known as the "transfer function". Figure 4.2 depicts three of the most commonly used activation functions in backpropagation MLPs:

Logistic function: $$f(x) = \frac{1}{1 + e^{-x}}$$

Hyperbolic tangent function: $$f(x) = \tan h\,(x/2) = \frac{1 - e^{-x}}{1 + e^{-x}}$$

Identity function: $$f(x) = x$$

Both the hyperbolic tangent and logistic functions approximate the signum and step function, respectively, and provide smooth, nonzero derivatives with respect to input signals. Sometimes these two activation functions are referred to as "squashing functions" since the inputs to these functions are squashed to the range [0,1] or [-1,1]. They are also called "sigmoidal functions" because their s-shaped curves exhibit smoothness and asymptotic properties.

Backpropagation MLPs are by far the most commonly used NN structures for applications in a wide range of areas, such as pattern recognition, signal processing, data compression and automatic control. Some of the well-known instances of applications include NETtalk (Sejnowski & Rosenberg, 1987), which trained an MLP to pronounce English text, Carnegie Mellon University's ALVINN (Pomerleau, 1991), which used an MLP for steering an autonomous vehicle; and optical character recognition (Sakinger, Boser, Bromley, Lecun & Jackel, 1992). In the following lines, we derive the backpropagation learning rule for MLPs using the logistic function.

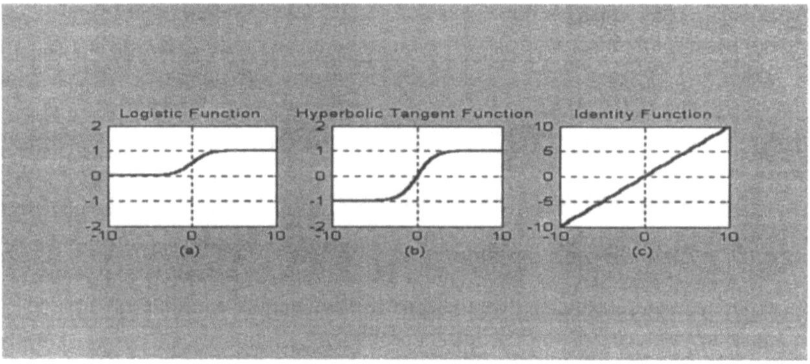

Figure 4.2 Activation functions for backpropagation MLPs: (a) logistic function; (b) hyperbolic function; (c) identity function.

The "net input" x of a node is defined as the weighted sum of the incoming signals plus a bias term. For instance, the net input and output of node j in Figure 4.3 are

$$x_j = \sum_i w_{ij}\, x_i + w_j \ , \qquad (4.10)$$
$$x_j = f(\, x_j\,) = \frac{1}{1 + \exp(-\, x_j\,)} \ ,$$

where x_i is the output of node i located in any one of the previous layers, w_{ij} is the weight associated with the link connecting nodes i and j, and w_j is the bias of node j. Since the weights w_{ij} are actually internal parameters associated with each node j, changing the weights of a node will alter the behavior of the node and in turn alter the behavior of the whole backpropagation MLP.

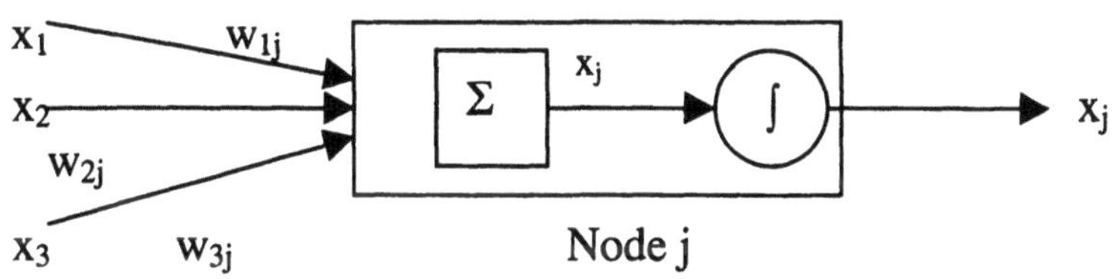

Figure 4.3 Node j of a backpropagation MLP.

Figure 4.4 shows a three-layer backpropagation MLP with three inputs to the input layer, three neurons in the hidden layer, and two output neurons in the output layer. For simplicity, this MLP will be referred to as a 3-3-2 network, corresponding to the number of nodes in each layer.

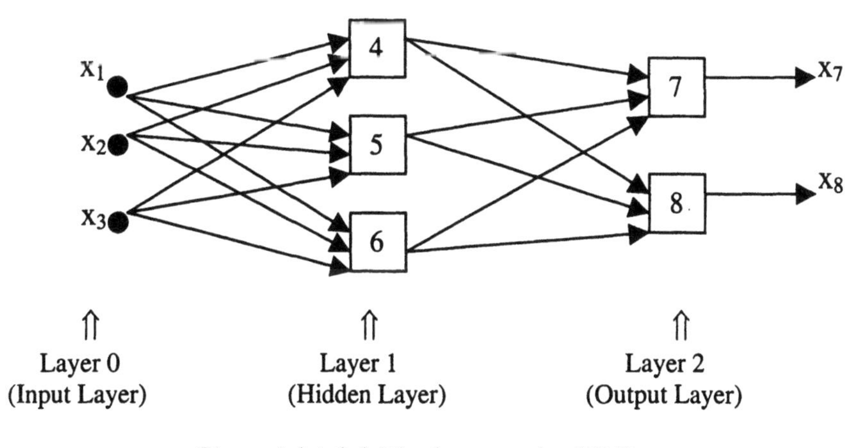

Figure 4.4 A 3-3-2 backpropagation MLP.

The "backward error propagation", also known as the "backpropagation" (BP) or the "generalized data rule" (GDR), is explained next. First, a squared error measure for the pth input-output pair is defined as

$$E_p = \sum_k (d_k - x_k)^2 \qquad (4.11)$$

where d_k is the desired output for node k, and x_k is the actual output for node k when the input part of the pth data pair presented. To find the gradient vector, an error term ε_i is defined as

$$\varepsilon_i = \frac{\partial^+ E_p}{\partial x_i} \qquad (4.12)$$

By the chain rule, the recursive formula for ε_i can be written as

$$\varepsilon_i = \begin{cases} -2(d_i - x_i) \dfrac{\partial x_i}{\partial x_i} = -2(d_i - x_i)\, x_i\, (1 - x_i) & \text{if node i is a output node} \\ \dfrac{\partial x_i}{\partial x_i} \sum_{j,i<j} \dfrac{\partial^+ E_p}{\partial x_j} \dfrac{\partial x_j}{\partial x_i} = x_i\, (1 - x_i) \sum_{j,i<j} \varepsilon_j\, w_{ij} & \text{otherwise} \end{cases} \qquad (4.13)$$

where w_{ij} is the connection weight from node i to j; and w_{ij} is zero if there is no direct connection. Then the weight update w_{ki} for on-line (pattern-by-pattern) learning is

$$\Delta w_{ki} = -\eta \frac{\partial^+ E_p}{\partial w_{ki}} = -\eta \frac{\partial^+ E_p}{\partial x_i} \frac{\partial x_i}{\partial w_{ki}} = -\eta\, \varepsilon_i\, x_k \qquad (4.14)$$

where η is a learning rate that affects the convergence speed and stability of the weights during learning.

For off-line (batch) learning, the connection weight w_{ki} is updated only after presentation of the entire data set, or only after an "epoch":

$$\Delta w_{ki} = -\eta \frac{\partial^+ E}{\partial w_{ki}} = -\eta \sum_p \frac{\partial^+ E_p}{\partial w_{ki}} \qquad (4.15)$$

or, in vector form,

$$\Delta \mathbf{w} = -\eta \frac{\partial^{+} E}{\partial \mathbf{w}} = -\eta \nabla_{\mathbf{w}} E \tag{4.16}$$

where $E = \sum_p E_p$. This corresponds to a way of using the true gradient direction based on the entire data set.

The approximation power of backpropagation MLPs has been explored by some researchers. Yet there is very little theoretical guidance for determining network size in terms of say, the number of hidden nodes and hidden layers it should contain. Cybenko (1989) showed that a backpropagation MLP, with one hidden layer and any fixed continuous sigmoidal non-linear function, can approximate any continuous function arbitrarily well on a compact set. When used as a binary-valued neural network with the step activation function, a backpropagation MLP with two hidden layers can form arbitrary complex decision regions to separate different classes, as Lippmann (1987) pointed out. For function approximation as well as data classification, two hidden layers may be required to learn a piecewise-continuous function (Masters, 1993).

Lets consider a simple example to illustrate the backpropagation learning algorithm. We will consider as training data the set of points distributed in the square $[-1, 1] \times [-1, 1]$, which are shown in Figure 4.5.

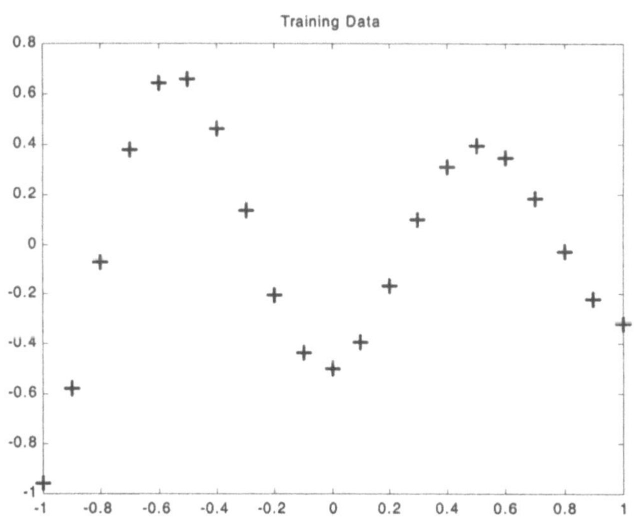

Figure 4.5 Training data for the backpropagation learning algorithm.

We show in Figure 4.6 the initial approximation of a three layer neural network with 10 neurons in the hidden layer and hyperbolic tangent activation functions. The initial approximation is quite bad because the initial weights of the network are generated randomly. After training with the backpropagation algorithm for

1000 epochs with arrive to the final approximation shown in Figure 4.7, which has a final sum of squares errors SSE = 0.0264283.

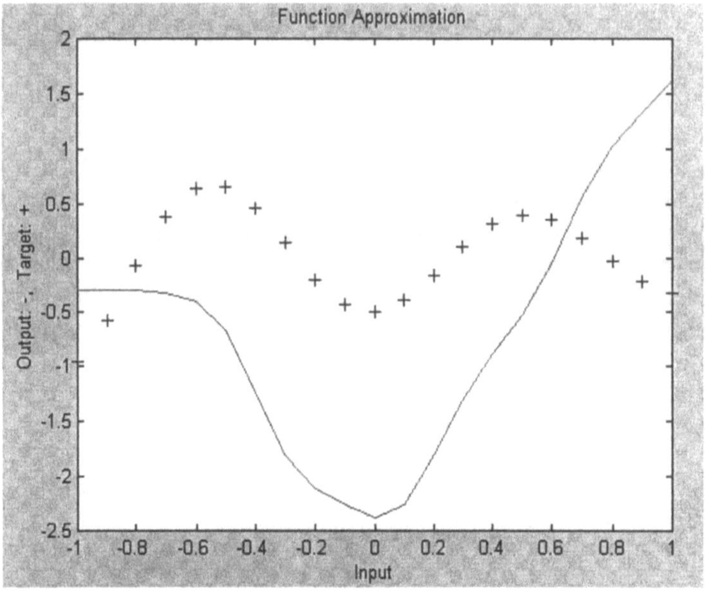

Figure 4.6 Initial approximation of the backpropagation learning algorithm.

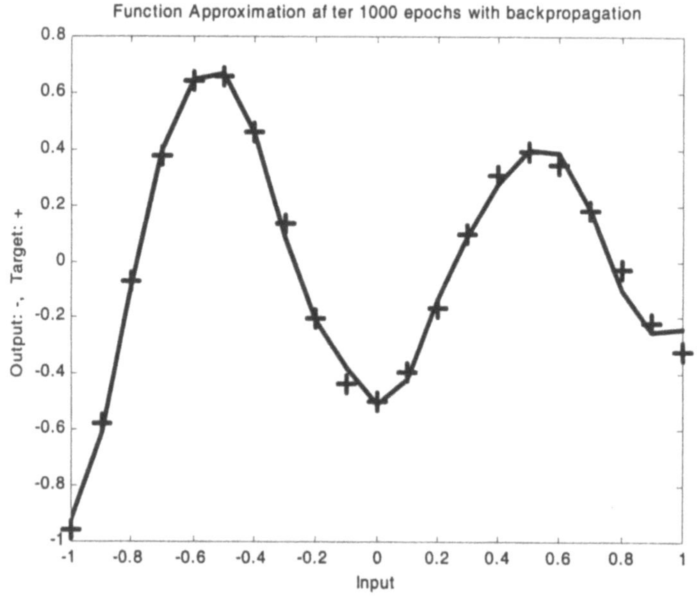

Figure 4.7 Final approximation backpropagation after 1000 epochs.

4.1.3 Methods for Speeding Up Backpropagation

One way for to speed up backpropagation is to use the so-called "momentum term" (Rumelhart, Hinton and Williams, 1986):

$$\Delta \mathbf{w} = \eta \nabla_{\mathbf{w}} E + \alpha \, \Delta \mathbf{w}_{prev} \quad (4.17)$$

where $\mathbf{w}_{prev}$ is the previous value of the weights, and the "momentum constant" α, in practice, is usually set to a value between 0.1 and 1. The addition of the momentum term smoothes weight updating and tends to resist erratic weight changes due to gradient noise or high spatial frequencies in the error surface. However, the use of momentum terms does not always seem to speed up training; it is more or less application dependent.

Another useful technique is normalized weight updating:

$$\Delta \mathbf{w} = - \kappa \, (\nabla_{\mathbf{w}} E) / (\|\nabla_{\mathbf{w}} E\|) \quad (4.18)$$

This causes the network's weight vector to move the same Euclidean distance in the weight space with each update, which allows control of the distance κ based on the history of error measures. Other strategies for speeding up backpropagation training include the quick-propagation algorithm (Fahlman, 1988), backpropagation with adaptive learning rate, and backpropagation with momentum and adaptive learning rate (Jang, Sun and Mizutani, 1997).

Lets consider a simple example to illustrate these methods. We will again use as training data the set of points shown in Figure 4.5. We will first apply the backpropagation with momemtum learning algorithm with the same parameters as before. We show in Figure 4.8 the initial function approximation and in Figure 4.9 the final function approximation achieved with backprogration with momentum.

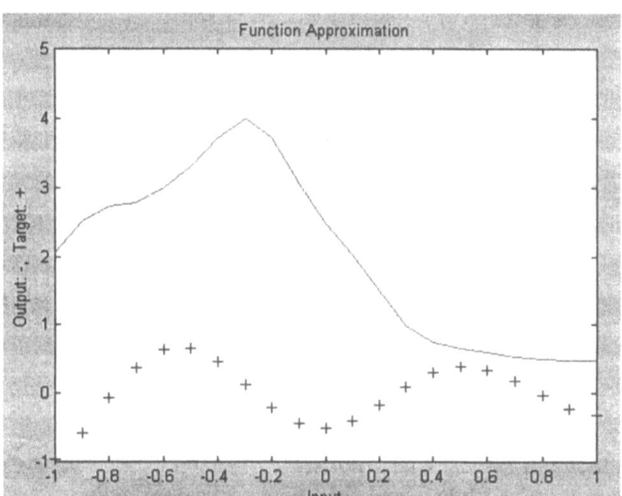

Figure 4.8 Initial function approximation with backpropagation with momentum.

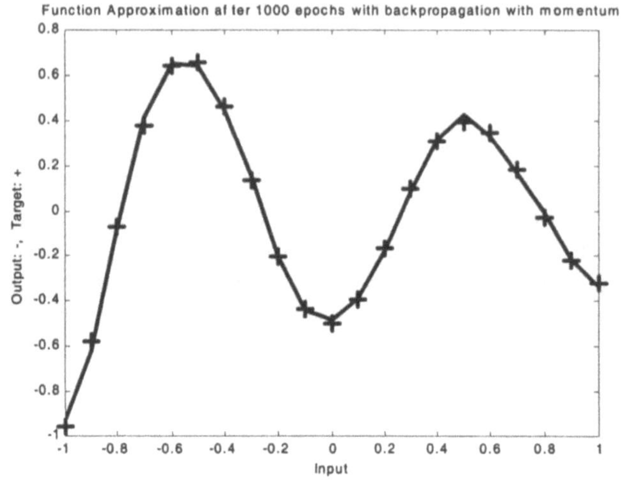

Figure 4.9 Final function approximation with backpropagation with momentum.

In this case, the final SSE is of 0.0082689, which is lower than the one obtained by simple backpropagation. As a consequence we are achieving a better final approximation with the the backpropagation with momentum.

Now we will consider the use of "backpropagation with momentum and adaptive learning rate". In this case, the learning rate is not fixed as in the previous methods, instead it is changed according to the error surface. We will again consider the training data of Figure 4.5. The initial function approximation is shown in Figure 4.10. The final function approximation is shown in Figure 4.11, which is achieved after 1000 epochs with the same network architecture as before.

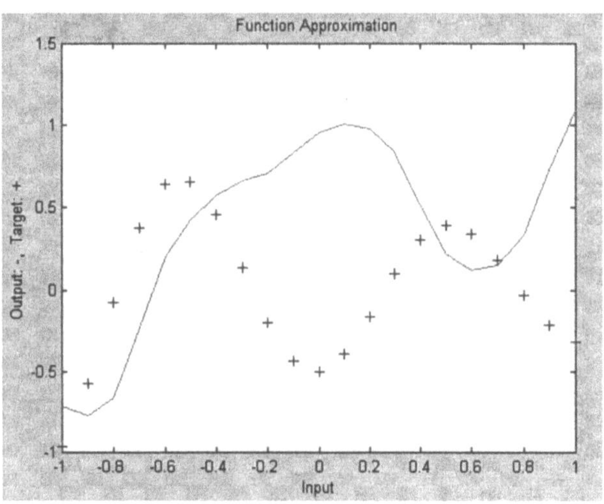

Figure 4.10 Initial approximation with backpropagation and adaptive learning rate.

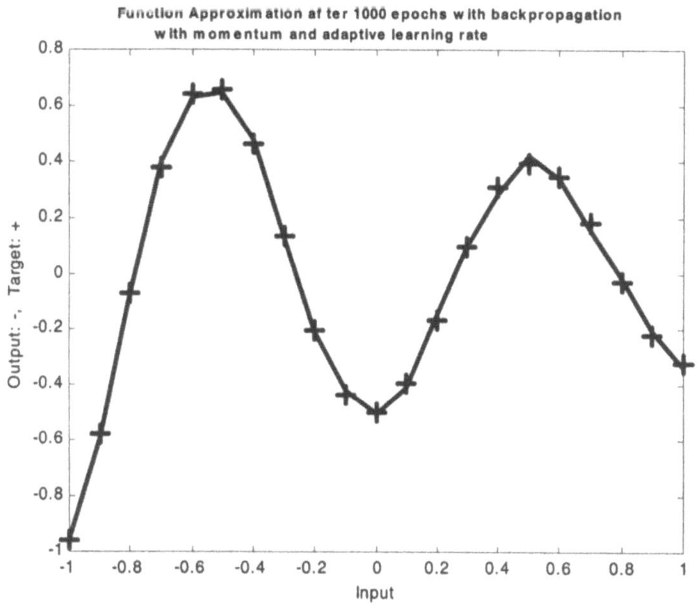

Figure 4.11 Final approximation with backpropagation and adaptive learning rate.

In this case, the final approximation achieved with the "backpropagation method with adaptive learning rate" is even better because the SSE is of only 0.0045014. This SSE is lower than the ones obtained previously with the other methods.

4.2 Radial Basis Function Networks

Locally tuned and overlapping receptive fields are well-known structures that have been studied in regions of the cerebral cortex, the visual cortex, and others. Drawing on knowledge of biological receptive fields, Moody and Darken (1989) proposed a network structure that employs local receptive fields to perform function mappings. Similar schemes have been proposed by Powell (1987) and many others in the areas of "interpolation" and "approximation theory"; these schemes are collectively call radial basis function approximations. Here we will call the neural network structure the "radial basis function network" or RBFN.

Figure 4.12 shows a schematic diagram of a RBFN with four receptive field units; the activation level of the *i*th receptive field unit (or hidden unit) is

$$w_i = R_i(\mathbf{x}) = R_i(\|\mathbf{x} - \mathbf{u}_i\| / \sigma_i), \; i = 1, 2, \ldots, H, \qquad (4.19)$$

where $\mathbf{x}$ is a multidimensional input vector, $\mathbf{u}_i$ is a vector with the same dimension as $\mathbf{x}$, H is the number of radial basis functions (or, equivalently, receptive field units), and $R_i(\ )$ is the *i*th radial basis function with a single maximum at the origin. There are no connection weights between the input layer and the hidden layer. Typically, $R_i(\ )$ is a Gaussian function

$$R_i(\mathbf{x}) = \exp [- (\| \mathbf{x} - \mathbf{u}_i \|^2)/ 2\sigma^2_i] \tag{4.20}$$

or a logistic function

$$R_i(\mathbf{x}) = 1 / [1 + \exp [(\| \mathbf{x} - \mathbf{u}_i \|^2)/ \sigma^2_i] \tag{4.21}$$

Thus, the activation level of radial basis function wi computed by the *i*th hidden unit is maximum when the input vector $\mathbf{x}$ is at the center $\mathbf{u}_i$ of that unit.

The output of an RBFN can be computed in two ways. In the simpler method, as shown in Figure 4.12, the final output is the weighted sum of the output value associated with each receptive field:

$$d(\mathbf{x}) = \Sigma^H{}_{i=1}\, c_i\, w_i = \Sigma^H{}_{i=1}\, c_i\, R_i(\mathbf{x}) \tag{4.22}$$

where c_i is the output value associated with the *i*th receptive field. We can also view c_i as the connection weight between the receptive field *i* and the ouput unit. A more complicated method for calculating the overall output is to take the weighted average of the output associated with each receptive field:

$$d(\mathbf{x}) = (\Sigma^H{}_{i=1}\, c_i\, w_i)/ (\Sigma^H{}_{i=1}\, w_i) \tag{4.23}$$

Weighted average has a higher degree of computational complexity, but it has the advantage that points in the areas of overlap between two or more receptive fields will have a well-interpolated overall output between the outputs of the overlapping receptive fields.

For representation purposes, if we change the radial basis function $R_i(x)$ in each node of layer 2 in Figure 4.12 to its normalized counterpart $R_i(x)/ \Sigma_i R_i(x)$, then the overall output is specified by Equation (4.23). A more explicit representation is the shown in Figure 4.13, where the division of the weighted sum $(\Sigma_{i=1}\, c_i\, w_i)$ by the activation total $(\Sigma_{i=1}\, w_i)$ is indicated in the division node in the last layer. Of course, similar figures can be drawn for two inputs or more in a RBFN network. We can appreciate from these figures the architecture of this type of neural networks. As a consequence we can see the difference between RBFN neural networks and MLP networks.

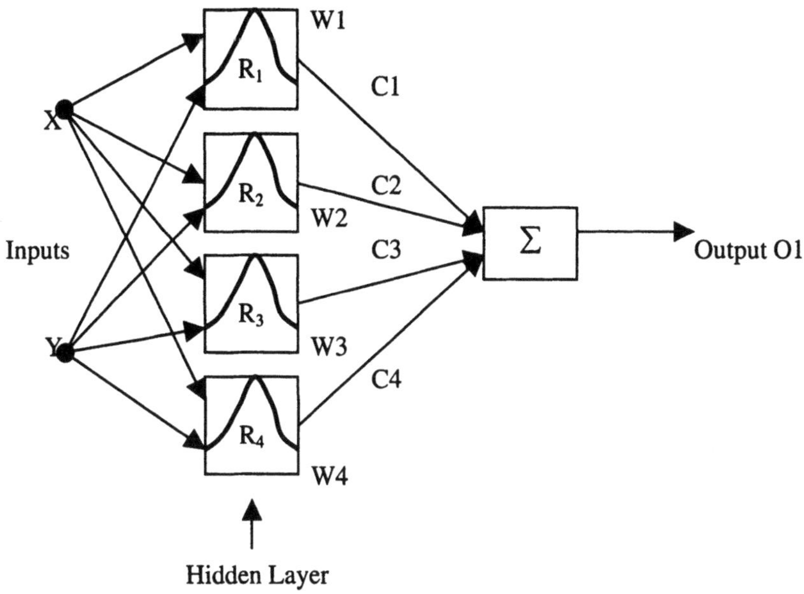

Figure 4.12 Single-output RBFN that uses weighted sum.

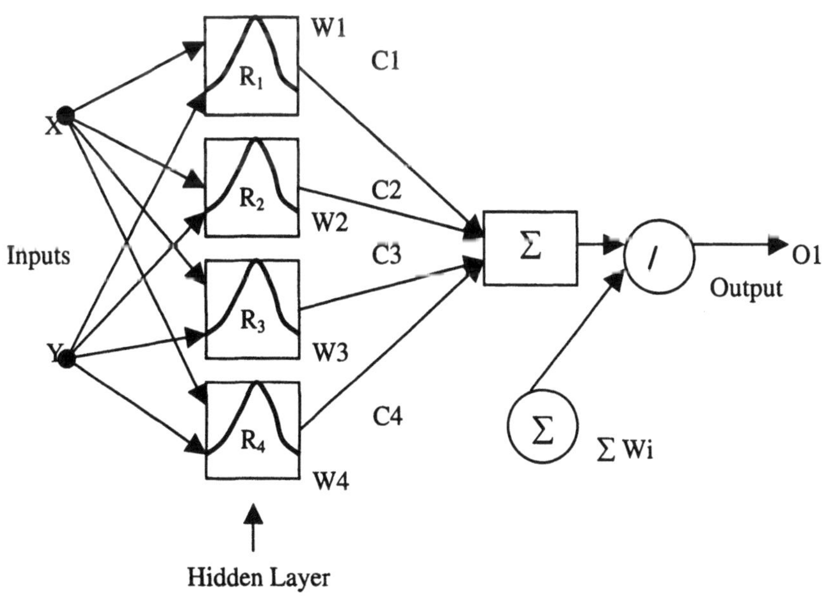

Figure 4.13 Single-output RBFN that uses weighted average.

Moody-Darken's RBFN may be extended by assigning a linear function to the output function of each receptive field-that is, making c_i a linear combination of the input variables plus a constant:

$$c_i = \mathbf{a_i}^T \mathbf{x} + b_i \tag{4.24}$$

where $\mathbf{a}_i$ is a parameter vector and b_i is a scalar parameter. An RBFN's approximation capacity may be further improved with supervised adjustments of the center and shape of the receptive field (or radial basis) functions (Lee & Kil, 1991). Several learning algorithms have been proposed to identify the parameters ($\mathbf{u}_i$, σ_i, and c_i) of an RBFN. Besides using a supervised learning scheme alone to update all modifiable parameters, a variety of sequential training algorithms for RBFNs have been reported. The receptive field functions are first fixed, and then the weights of the output layer are adjusted. Several schemes have been proposed to determine the center positions ($\mathbf{u}_i$) of the receptive field functions. Lowe (1989) proposed a way to determine the centers based on standard deviations of the training data. Moody and Darken (1989) selected the centers $\mathbf{u}_i$ by means of data clustering techniques that assume that similar input vectors produce similar outputs; σ_i's are then obtained heuristically by taking the average distance to the several nearest neighbors of u_i's. Once the non-linear parameters are fixed and the receptive fields are frozen, the linear parameters (i.e., the weights of the output layer) can be updated using either the least squares method or the gradient method.

Using Equation (4.24), extended RBFN response is identical to the response produced by the first-order Sugeno (type-1) fuzzy inference system described in Chapter 2, provided that the membership functions, the radial basis functions, and certain operators are chosen correctly. While the RBFN consists of radial basis functions, the Sugeno fuzzy system contains a certain number of membership functions. Although the fuzzy system and the RBFN were developed on different bases, they are essentially rooted in the same grounds. Just as the RBFN enjoys quick convergence, the fuzzy system can evolve to recognize some features in a training data set.

Assuming that there is no noise in the training data set, we need to estimate a function d(.) that yields exact desired outputs for all training data. This task is usually called an "interpolation" problem, and the resulting function d(.) should pass through all of the training data points. When we use an RBFN with the same number of basis functions as we have training patterns, we have a so-called "interpolation RBFN", where each neuron in the hidden layer responds to one particular training input pattern.

Lets consider application of the RBFN network to the same example of Figure 4.5. We will use a two layer RBFN network with 3 neurons in the hidden layer and weighted sum to calculate the output. We show in Figure 4.14 the Gaussian radial basis function used in the network. Figure 4.15 illustrates the application of weighted sum to achieve the approximation of the training data. Figure 4.16 shows the final approximation achieved with the RBFN network, which is very good. The final SSE is of only 0.002, which is smaller than the one obtained by any of the previous methods. We can conclude that the RBFN network gives the best approximation to the training data of Figure 4.5.

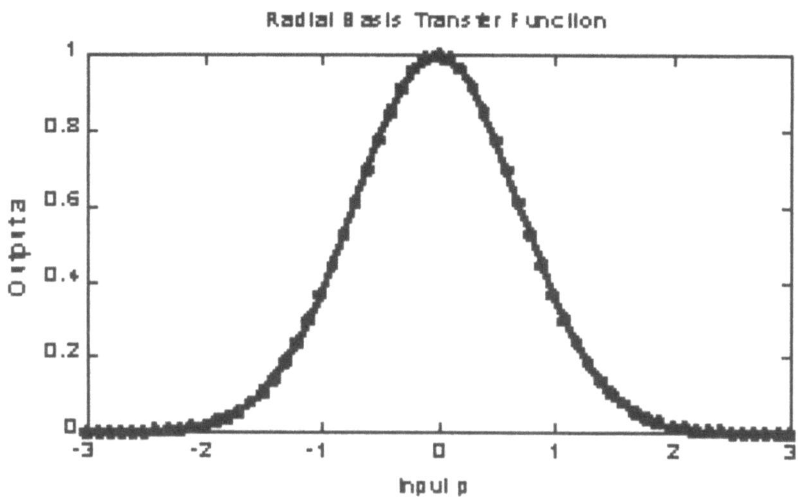

Figure 4.14 Gaussian radial basis function.

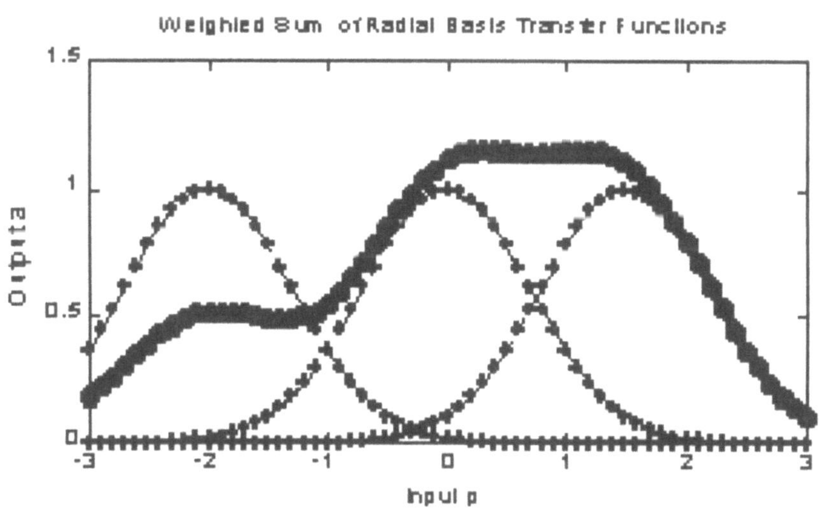

Figure 4.15 Weighted sum of the three Gaussian functions of the RBFN network.

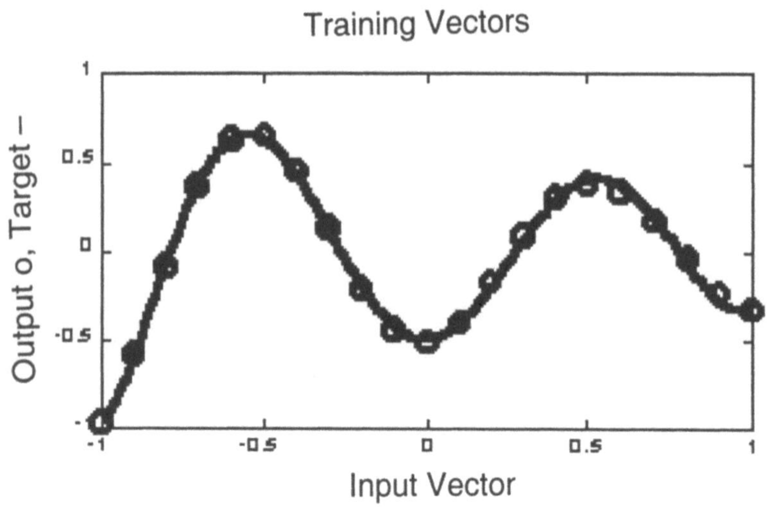

Figure 4.16 Final function approximation achieved with the RBFN network.

4.3 Adaptive Neuro-Fuzzy Inference Systems

In this section, we describe a class of adaptive networks that are functionally equivalent to fuzzy inference systems (Kosko, 1992). The architecture is referred to as ANFIS, which stands for "adaptive network-based fuzzy inference system". We describe how to decompose the parameter set to facilitate the hybrid learning rule for ANFIS architectures representing both the Sugeno and Tsukamoto fuzzy models.

4.3.1 ANFIS Architecture

A fuzzy inference system consists of three conceptual components: a fuzzy rule base, which contains a set of fuzzy if-then rules; a database, which defines the membership functions used in the fuzzy rules; and a reasoning mechanism, which performs the inference procedure upon the rules to derive a reasonable output or conclusion (Kandel, 1992). For simplicity, we assume that the fuzzy inference system under consideration has two inputs x and y and one output z. For a first-order Sugeno fuzzy model (Sugeno & Kang, 1988), a common rule set with two fuzzy if-then rules is the following:

Rule 1: If x is A_1 and y is B_1, then $f_1 = p_1x + q_1y + r_1$,

Rule 2: If x is A_2 and y is B_2, then $f_2 = p_2x + q_2y + r_2$,

Figure 4.17 (a) illustrates the reasoning mechanism for this Sugeno model; the corresponding equivalent ANFIS architecture is as shown in Figure 4.17 (b), where nodes of the same layer have similar functions, as described next. (Here we denote the output of the ith node in layer l as $0_{l,i}$).

Layer 1: Every node i in this layer is an adaptive node with a node function

$$0_{l,i} = \mu_{Ai}(x), \text{ for } i = 1, 2,$$
$$0_{l,i} = \mu_{Bi-2}(y), \text{ for } i = 3, 4, \qquad (4.25)$$

where x (or y) is the input to node i and A_i (or B_{i-2}) is a linguistic label (such as "small" or "large") associated with this node. In other words, $0_{l,i}$ is the membership grade of a fuzzy set A and it specifies the degree to which the given input x (or y) satisfies the quantifier A. Here the membership function for A can be any appropriate parameterized membership function, such as the generalized bell function:

$$\mu_A(x) = \frac{1}{1 + |(x-c_i)/a_i|^{2bi}} \qquad (4.26)$$

where $\{a_i, b_i, c_i\}$ is the parameter set. As the values of these parameters change, the bell-shaped function varies accordingly, thus exhibiting various forms of membership functions for a fuzzy set A. Parameters in this layer are referred to as "premise parameters".

Layer 2: Every node in this layer is a fixed node labeled Π, whose output is the product of all incoming signals:

$$0_{2,i} = w_i = \mu_{Ai}(x)\, \mu_{Bi}(y), \quad i = 1, 2. \qquad (4.27)$$

Each node output represents the firing strength of a fuzzy rule.

Layer 3: Every node in this layer is a fixed node labeled N. The ith node calculates the ratio of the ith rule's firing strength to the sum of all rules' firing strengths:

$$0_{3,i} = w_i = w_i / (w_1 + w_2), \quad i = 1,2. \qquad (4.28)$$

For convenience, outputs of this layer are called "normalized firing strengths".

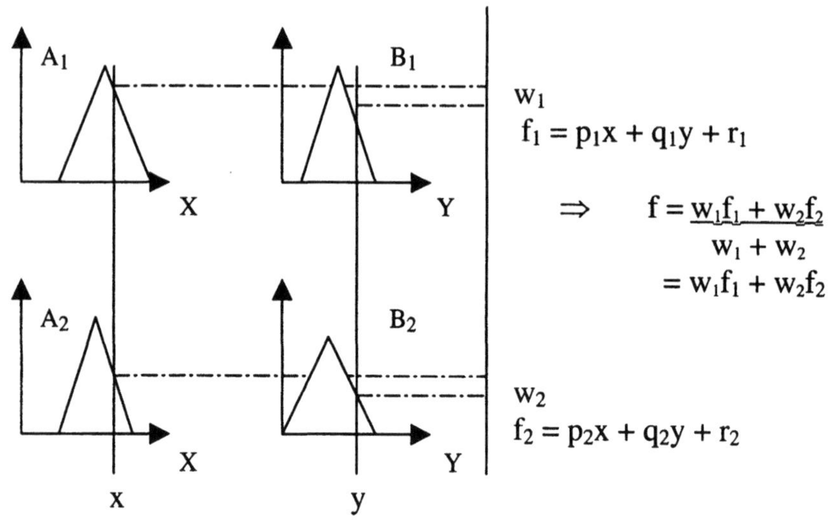

(a)

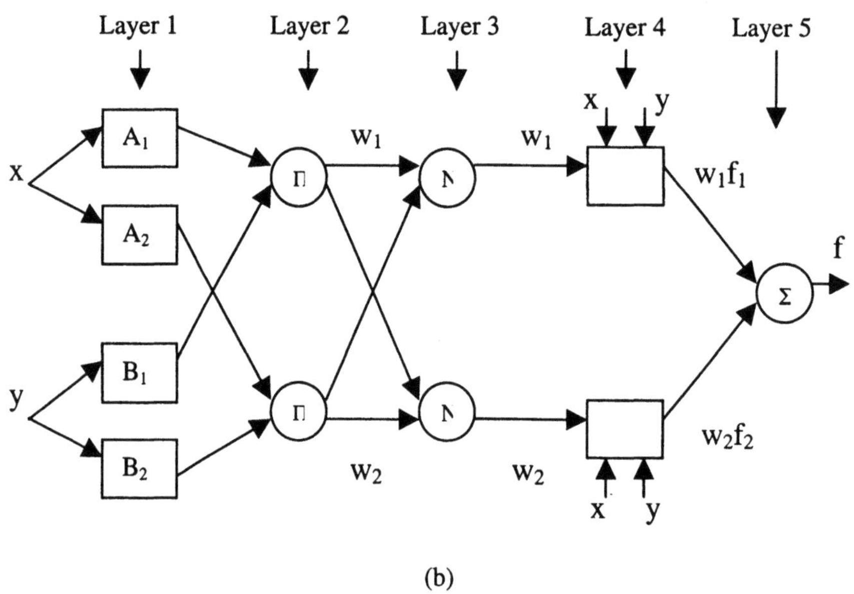

(b)

Figure 4.17 (a) A two-input Sugeno fuzzy model with 2 rules; (b) equivalent ANFIS architecture (adaptive nodes shown with a square and fixed nodes with a circle).

Layer 4: Every node i in this layer is an adaptive node with a node function

$$O_{4,i} = w_i f_i = w_i (p_i x + q_i y + r_i) \quad , \tag{4.29}$$

where w_i is a normalized firing strength from layer 3 and $\{p_i , q_i , r_i \}$ is the parameter set of this node. Parameters in this layer are referred to as "consequent parameters".

Layer 5: The single node in this layer is a fixed node labeled Σ, which computes the overall output as the summation of all incoming signals:

$$\text{overall output} = O_{5,i} = \sum_i w_i f_i = \frac{\sum_i w_i f_i}{\sum_i w_i} \tag{4.30}$$

Thus we have constructed an adaptive network that is functionally equivalent to a Sugeno fuzzy model. We can note that the structure of this adaptive network is not unique; we can combine layers 3 and 4 to obtain an equivalent network with only four layers. In the extreme case, we can even shrink the whole network into a single adaptive node with the same parameter set. Obviously, the assignment of node functions and the network configuration are arbitrary, as long as each node and each layer perform meaningful and modular functionalities.

The extension from Sugeno ANFIS to Tsukamoto ANFIS is straightforward, as shown in Figure 4.18, where the output of each rule (f_i, i = 1, 2) is induced jointly by a consequent membership function and a firing strength.

4.3.2 Learning algorithm

From the ANFIS architecture shown in Figure 4.7 (b), we observe that when the values of the premise parameters are fixed, the overall output can be expressed as a linear combination of the consequent parameters. Mathematically, the output f in Figure 4.7 (b) can be written as

$$\begin{aligned} f &= \frac{w_1}{w_1 + w_2} f_1 + \frac{w_2}{w_1 + w_2} f_2 \\ &= w_1 (p_1 x + q_1 y + r_1) + w_2 (p_2 x + q_2 y + r_2) \\ &= (w_1 x)\, p_1 + (w_1 y)\, q_1 + (w_1)\, r_1 + (w_2 x)\, p_2 + (w_2 y)\, q_2 + (w_2)\, r_2 \end{aligned} \tag{4.31}$$

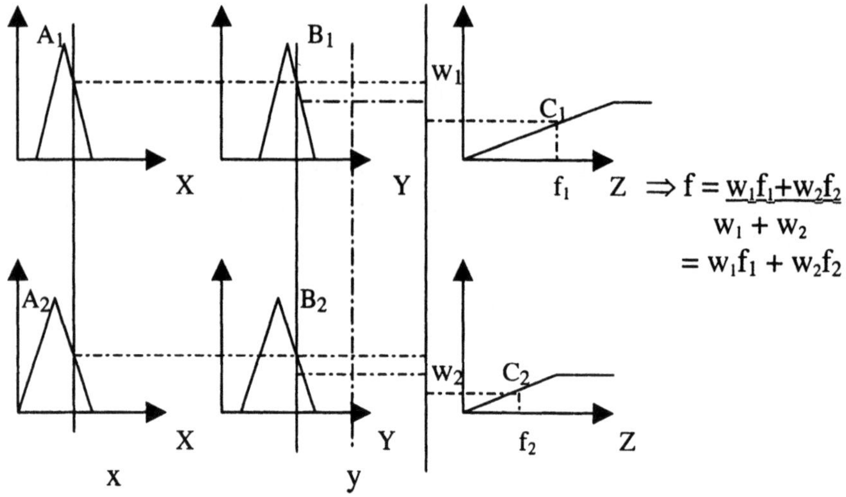

(a)

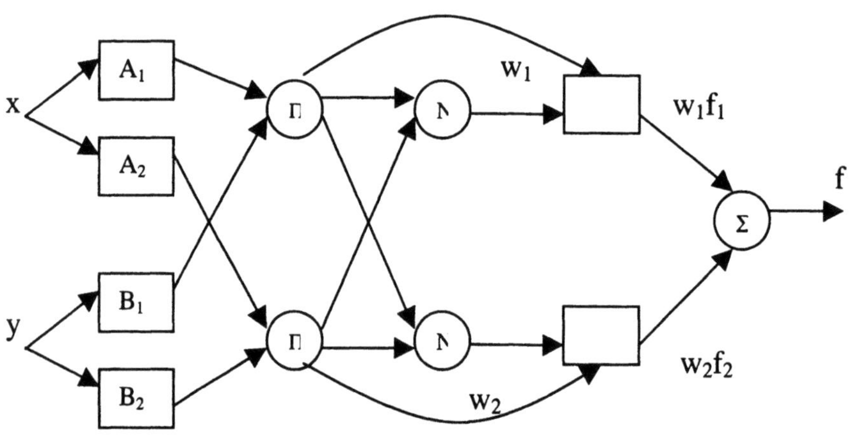

(b)

Figure 4.18 (a) A two-input Tsukamoto fuzzy model with two rules; (b) equivalent ANFIS architecture.

which is linear in the consequent parameters p_1, q_1, r_1, p_2, q_2, and r_2. From this observation, we can use a hybrid learning algorithm for parameter estimation in this kind of models (Jang, 1993). More specifically, in the forward pass of the hybrid learning algorithm, node outputs go forward until layer 4 and the consequent parameters are identified by the least-squares method. In the backward pass, the error signals propagate backward and the premise parameters are updated by gradient descent.

It has been shown (Jang, 1993) that the consequent parameters identified in this manner are optimal under the condition that the premise parameters are fixed. Accordingly, the hybrid approach converges much faster since it reduces the search space dimensions of the original pure backpropagation method. For Tsukamoto ANFIS, this can be achieved if the membership function on the consequent part of each rule is replaced by a piecewise linear approximation with two consequent parameters.

As we discussed earlier, under certain minor conditions, an RBFN is functionally equivalent to a fuzzy system, and thus to ANFIS. This functional equivalence provides a shortcut for better understanding both ANFIS and RBFNs in the sense that development in either literature cross-fertilize the other (Jang, Sun, and Mizutani, 1997).

Finally, we have to mention that it has been shown that the ANFIS methodology can be viewed as universal approximator (Jang, Sun, and Mizutani, 1997). More specifically, it has been shown that when the number of rules is not restricted, a zero-order Sugeno model has unlimited approximation power for matching any non-linear function arbitrarily well on a compact set. This fact is intuitively reasonable. However, the mathematical proof can be made by showing that ANFIS satisfies the well-known Stone-Weierstrass theorem (Kantorovich and Akilov, 1982).

We will now show a simple example to illustrate the ANFIS methodology. We will use as training data the set of points shown in Figure 4.5. We will use a network of 20 nodes, 4 rules, 4 Gaussian membership functions, and 16 unknown parameters. The complete network is shown in Figure 4.19, in which we can clearly see all the details mentioned above. We have to mention that the ANFIS methodology is been used here to obtain a first order Sugeno model. In Figure 4.20, we can appreciate the rate of convergence of ANFIS as the error is plotted against the number of epochs. From this figure it is clear that ANFIS can achieve a comparable error (with the previous methods in this Chapter) in only 20 epochs, which is a lot less than the 1000 epochs required by the networks presented before (for the same example). In Figure 4.21 we can see the final function approximation achieved by the ANFIS method, which is very good. In Figure 4.22 we show the non-linear surface of the final fuzzy system obtained by ANFIS. In Figure 4.23, we show the use of ANFIS with specific values. In this case, the "rule viewer" of the Fuzzy Logic Toolbox of MATLAB is used to obtain these results. Finally, we show in Figure 4.24 the membership functions for the input variable of ANFIS.

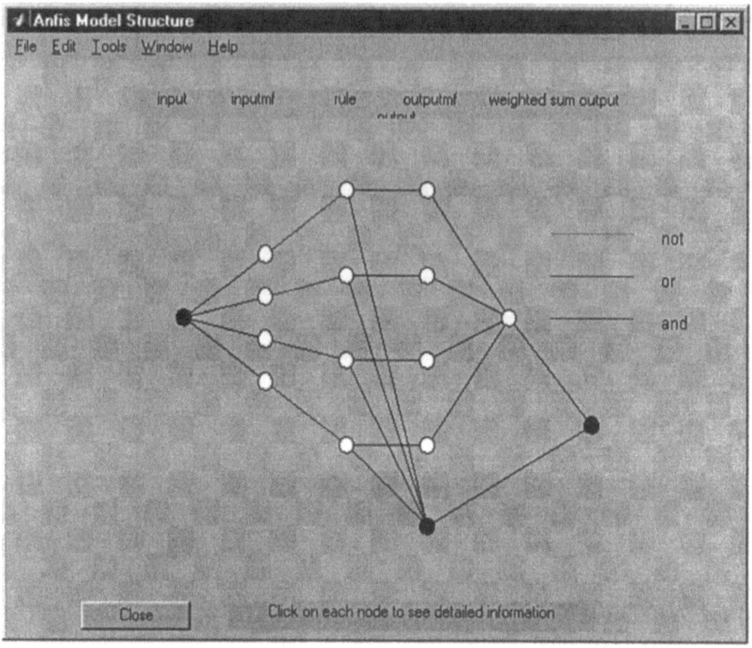

Figure 4.19 Architecture of network for the ANFIS method.

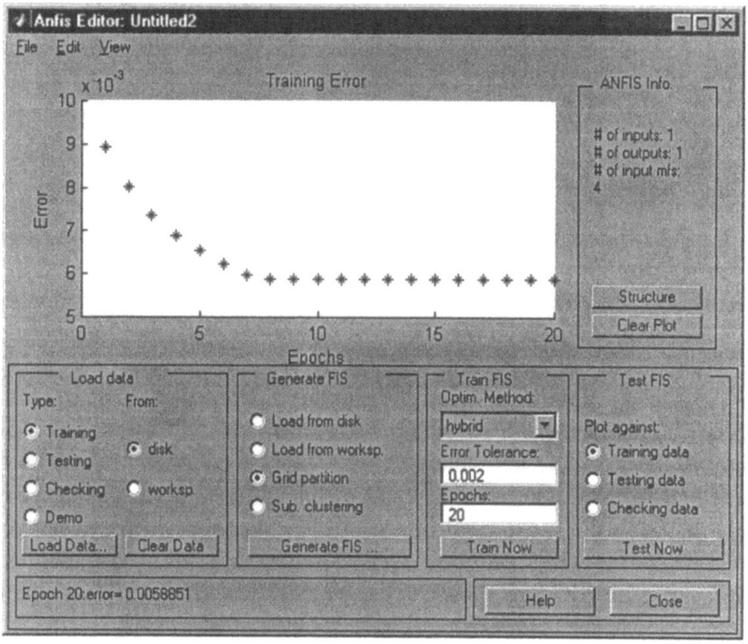

Figure 4.20 Convergence of ANFIS (final SSE = 0.0058851).

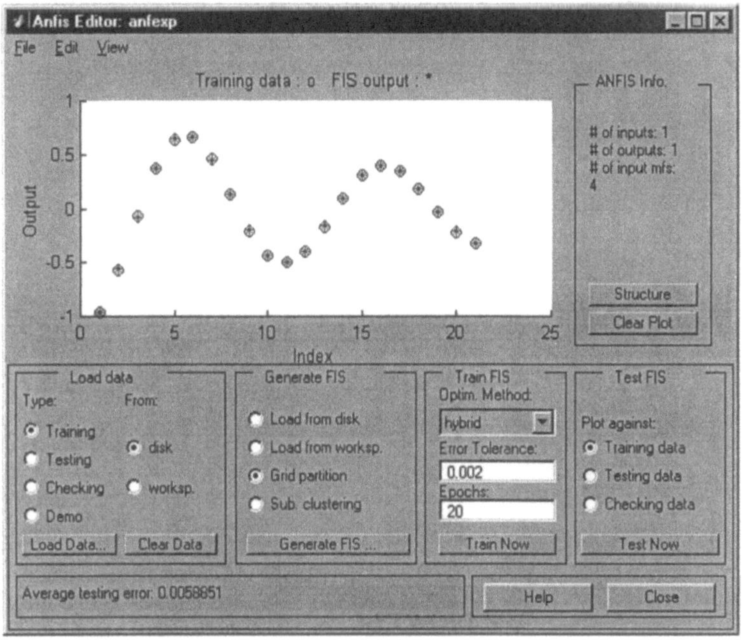

Figure 4.21 Final function approximation achieved by the ANFIS method.

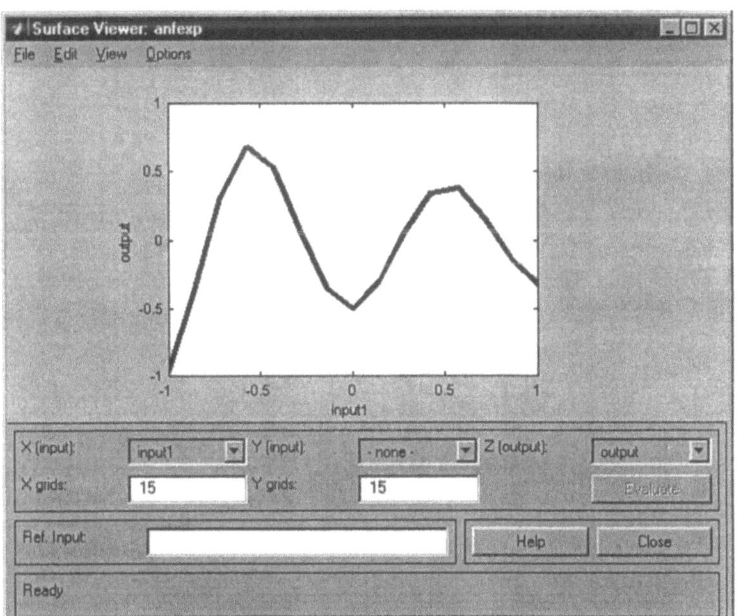

Figure 4.22 Non-linear surface obtained by the ANFIS method.

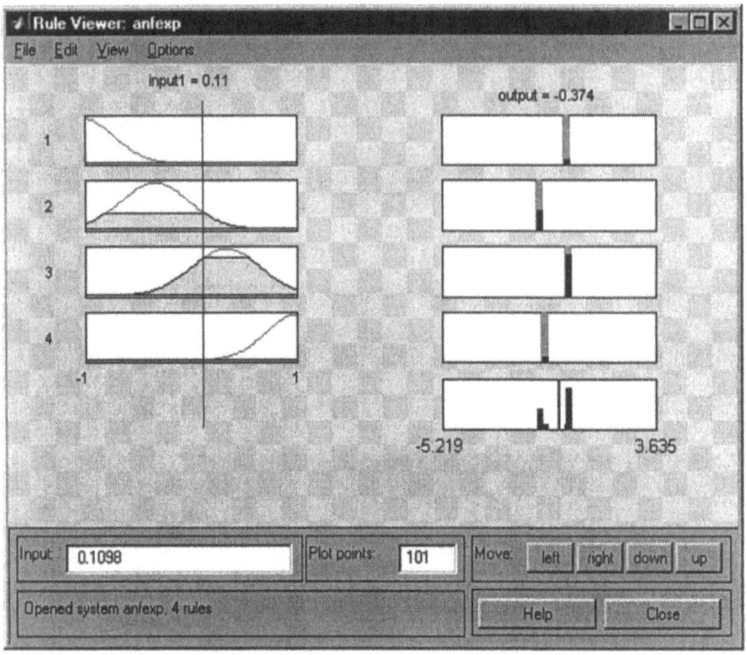

Figure 4.23 Use of ANFIS to calculate specific values with the rule viewer.

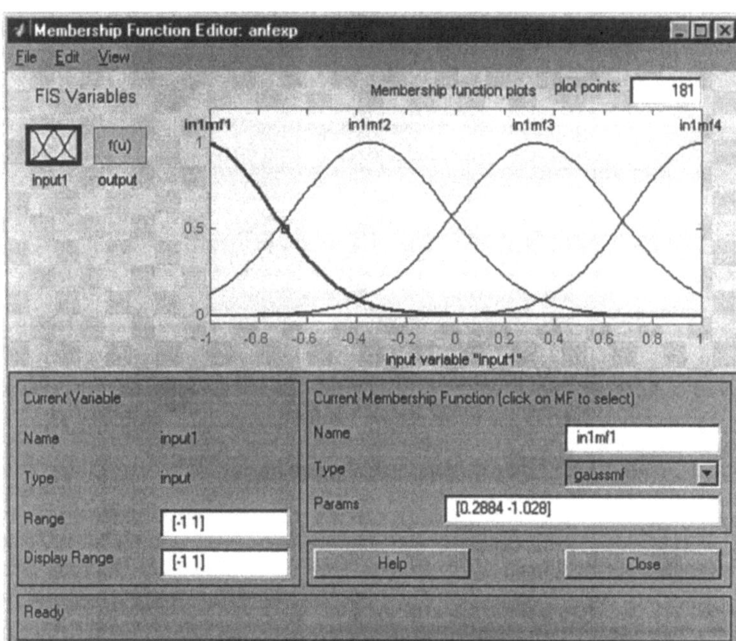

Figure 4.24 Membership functions for the input variable of ANFIS.

Finally, we have to say that the use of the ANFIS methodology is facilitated in the MATLAB programming language because it is already available in the Fuzzy Logic Toolbox. For this reason, all of the results shown before were obtained very easily using this tool of MATLAB.

4.4 Summary

In this chapter, we have presented the main ideas underlying supervised neural networks and the application of this powerful computational theory to general problems in function approximation. We have discussed in some detail the backpropagation learning algorithm for feedforward networks, radial basis function neural networks, and the integration of fuzzy logic techniques to neural networks to form powerful adaptive neuro-fuzzy inference systems. In the following chapters, we will show how supervised neural network techniques (in conjunction with other techniques) can be applied to solve real world complex problems in intelligent manufacturing. This chapter will serve as a basis for the new hybrid intelligent methods that will be described in the chapters at the end of this book.

Chapter 5

Unsupervised Learning Neural Networks

This chapter introduces the basic concepts and notation of unsupervised learning neural networks. Unsupervised networks are useful for analyzing data without having the desired outputs; in this case, the neural networks evolve to capture density characteristics of a data phase. We will describe in some detail competitive learning networks, Kohonen self-organizing networks, learning vector quantization, and Hopfield networks. We will also show some examples of these networks to illustrate their possible application in solving real-world problems.

When no external teacher or critic's instruction is available, only input vectors can be used for learning. Such an approach is learning without supervision, or what is commonly referred to as unsupervised learning. An unsupervised learning system evolves to extract features or regularities in presented patterns, without being told what outputs or classes associated with the input patterns are desired. In other words, the learning system detects or categorizes presistent features without any feedback from the environment. Thus unsupervised learning is frequently employed for data clustering, feature extraction, and similarity detection.

Unsupervised learning Neural Networks attempt to learn to respond to different input patterns with different parts of the network. The network is often trained to strengthen firing to respond to frequently occuring patterns, thereby leading to the so-called synonym probability estimators. In this manner, the network develops certain internal representations for encoding input patterns.

5.1 Competitive Learning Networks

With no available information regarding the desired outputs, unsupervised learning networks update weights only on the basis of the input patterns. The competitive learning network is a popular scheme to achieve this type of unsupervised data clustering or classification. Figure 5.1 presents an example of competitive learning network. All input units i are connected to all output units j with weight w_{ij}. The number of inputs is the input dimension, while the number of outputs is equal to the number of clusters that the data are to be divided into. A cluster center's position is specified by the weight vector connected to the corresponding output unit. For the simple network in Figure 5.1, the three-dimensional input data are divided into five clusters, and the cluster centers, denoted as the weights, are updated via the competitive learning rule.

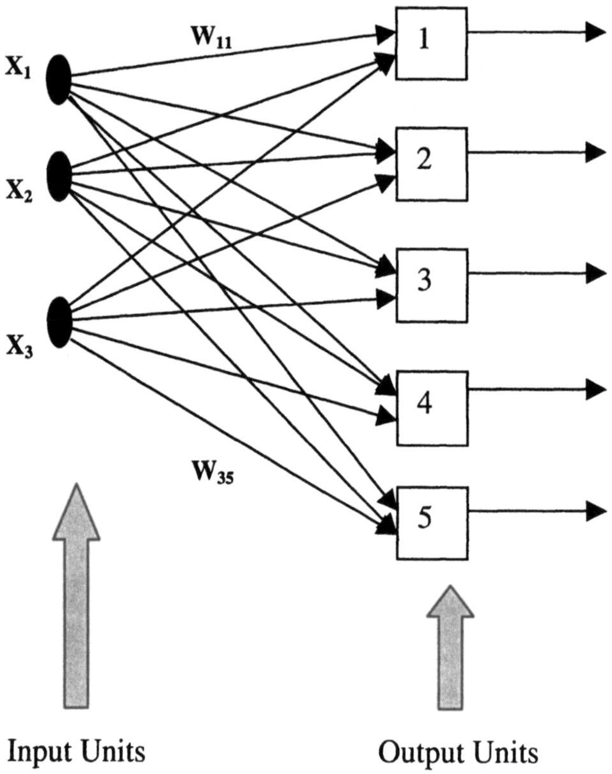

Figure 5.1 Competitive Learning Network.

The input vector $X = [x_1, x_2, x_3]^T$ and the weight vector $W_j = [w_{1j}, w_{2j}, w_{3j}]^T$ for an output unit j are generally assumed to be normalized to unit length. The activation value a_j of output unit j is then calculated by the inner product of the input and weight vectors:

$$a_j = \sum_{i=1}^{3} x_i w_{ij} = X^T W_j = W^T{}_J X \tag{5.1}$$

Next, the output unit with the highest activation must be selected for further processing, which is what is implied by competitive. Assuming that output unit k has the maximal activation, the weights leading to this unit are updated according to the competitive or the so-called winner-take-all learning rule:

$$w_k(t+1) = \frac{w_k(t) + \eta(x(t) - w_k(t))}{\|w_k(t) + \eta(x(t) - w_k(t))\|} \tag{5.2}$$

The preceding weight update formula includes a normalization operation to ensure that the updated weight is always of unit length. Notably, only the weights at the winner output unit k are updated; all other weights remain unchanged.

The update formula in Equation (5.2) implements a sequential scheme for finding the cluster centers of a data set of which the entries are of unit length. When an input x is presented to the network, the weight vector closest to x rotates toward it. Consequently, weight vectors move toward those areas where most inputs appear and, eventually, the weight vectors become the cluster centers for the data set. In Figure 5.2 we show this dynamic process.

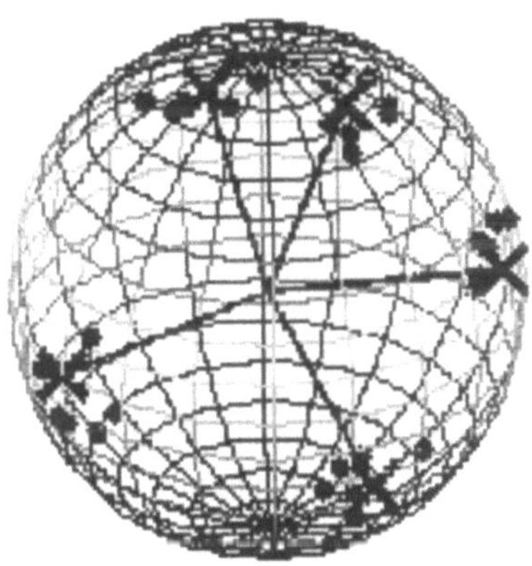

Figure 5.2 Competitive learning with unit-length vectors. The dots represent the input vectors and the crosses denote the weight vectors for the five output units in Figure 5.1. As the learning continues, the five weight vectors rotate toward the centers of the five input clusters.

Using the Euclidean distance as a dissimilarity measure is a more general scheme of competitive learning, in which the activation of output unit j is

$$a_j = \left\{ \sum_{i=1}^{3} (x_i - w_{ij})^2 \right\}^{0.5} = \| x - w_j \| \tag{5.3}$$

The weights of the output unit with the smallest activation are updated according to

$$w_k(t+1) = w_k(t) + \eta(x(t) - w_k(t)) \qquad (5.4)$$

In the Equation (5.4), the wining unit's weights shift toward the input x. In this case, neither the data nor the weights must be of unit length.

A competitive learning network performs an on-line clustering process on the input patterns. When the process is complete, the input data are divided into disjoint clusters such that similarities between individuals in the same cluster are larger than those in different clusters. Here two metrics of similarity are introduced: the similarity measure of inner product in Equation (5.1) and the dissimilarity measure of the Euclidean distance in Equation (5.3). Obviously, other metrics can be used instead, and different selections lead to different clustering results. When the Euclidean distance is adopted, it can be proved that the update formula in Equation (5.4) is actually an on-line version of gradient descent that minimizes the following objection function:

$$E = \sum_{p} \|w_{f(xp)} - x_p\|^2 \qquad (5.5)$$

where $f(x_p)$ is the wining neuron when input x_p is presented and $w_{f(xp)}$ is the center of the class where xp belongs to.

A large family of batch-mode (or off-line) clustering algorithms can be used to find cluster centers that minimize Equation (5.5) by example K-means clustering algorithm (Ruspini, 1982). A limitation of competitive learning is that some of the weight vectors that are initialized to random values may be far from any input vector and, subsequently, it never gets updated. Such a situation can be prevented by initializing the weights to samples from the input data itself, thereby ensuring that all of the weights get updated when all the input patterns are presented. An alternative would be to update the weights of both the winning and losing units, but use a significantly smaller learning rate h for the losers; this is commonly referred to as leaky learning (Rumelhart & Zipser, 1986).

Dynamically changing the learning rate η in the weight update formula of Equation (5.2) or (5.4) is generally desired. An initial large value of η explores the data space widely; later on, a progressively smaller value refines the weights. The operation is similar to the cooling schedule of simulated annealing, as introduced in Chapter 6. Therefore, one of the following formulas for η is commonly used:

$$\begin{cases} \eta(t) = \eta_o\, e^{-\alpha t}, & \text{with } \alpha > 0, \text{ or} \\ \eta(t) = \eta_o\, t^{-\alpha}, & \text{with } \alpha \leq 0, \text{ or} \\ \eta(t) = \eta_o(1-\alpha t), & \text{with } 0 < \alpha < (\max\{t\})^{-1} \end{cases}$$

Competitive learning lacks the capability to add new clusters when deemed necessary. Moreover, if the learning rate η is a constant, competitive learning does not guarantee stability in forming clusters; the winning unit that responds to a particular pattern may continue changing during training. On the other hand, η, if decreasing with time, may become too small to update cluster centers when new data of a different probability nature are presented. Carpenter and Grossberg referred to such an occurrence as the stability-plasticity dilemma, which is common in designing intelligent learning systems (Carpenter & Grossberg, 1988). In general, a learning system should be plastic, or adaptive in reacting to changing environments; meanwhile, it should be stable to preserve knowledge acquired previously.

If the output units of a competitive learning network are arranged in a geometric manner (such as in a one-dimensional vector or two-dimensional array), then we can update the weights of the winners as well as the neighboring losers. Such a capability corresponds to the notion of Kohonen feature maps, as discussed in the next section.

We illustrate the use of competitive learning networks with a simple example. In Figure 5.3 we show the initial data randomly distributed within the rectangle [-0.2, 1.2]× [0, 1]. We can also see at the center the initial weights of the network (the crosses indicate the weights). We did specified a-priori the number of clusters as eight. After training the competitive network for 500 epochs we obtained the result shown in Figure 5.4. In this figure we can see how the weights are arranged in their final positions, which are the centers of the clusters.

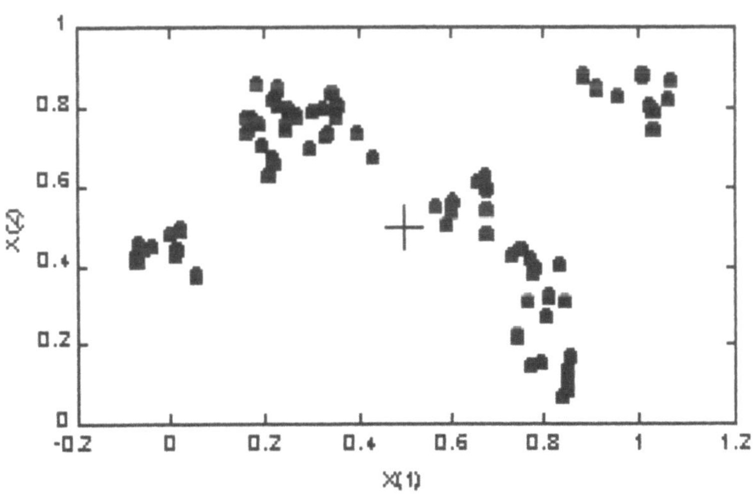

Figure 5.3 Initial randomly distributed data and weights for the competitive network.

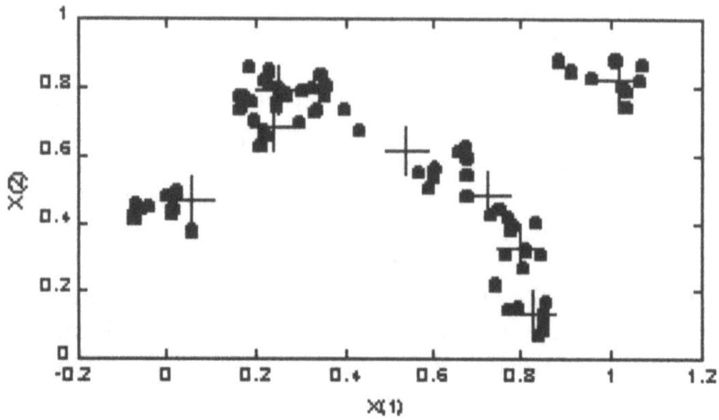

Figure 5.4 Final distribution of the 8 clusters after training for 500 epochs.

5.2 Kohonen Self-Organizing Networks

Kohonen self-organizing networks, also known as Kohonen feature maps or topology-preserving maps, are another competition-based network paradigm for data clustering (Kohonen, 1982, 1984). Networks of this type impose a neighborhood constraint on the output units, such that a certain topological property in the input data is reflected in the output units' weights.

In Figure 5.5 we show a relatively simple Kohonen self-organizing network with 2 inputs and 49 outputs. The learning procedure of Kohonen feature maps is similar to that of competitive learning networks. That is, a similarity (dissimilarity) measure is selected and the winning unit is considered to be the one with the largest (smallest) activation.

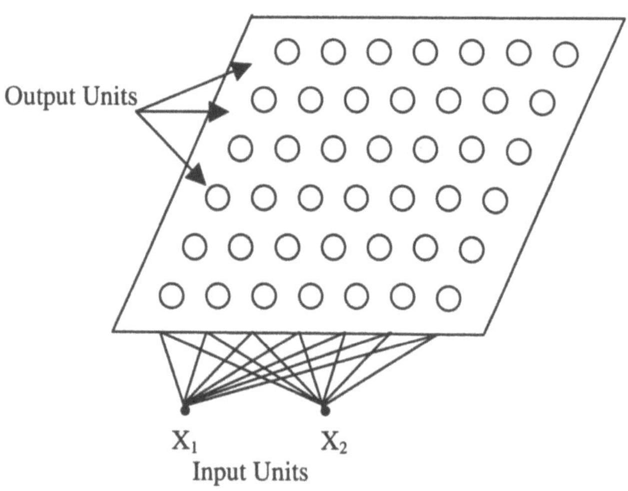

Figure 5.5 A Kohonen self-organizing network with 2 input and 49 output units.

For Kohonen feature maps, however, we update not only the winning unit's weights but also all of the weights in a neighborhood around the wining units. The neighborhood's size generally decreases slowly with each iteration, as indicated in Figure 5.6.

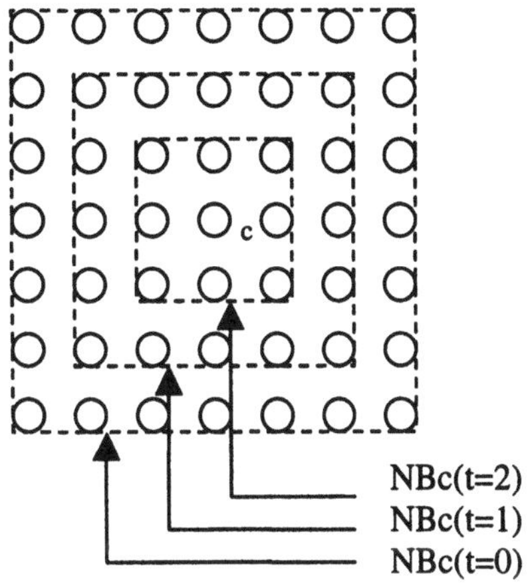

Figure 5.6 In Kohonen self-organizing network the size of a neighborhood around a winning unit decreases gradually with each iteration.

A sequential description of how to train a Kohonen self-organizing network is as follows:

Step1: Select the wining output unit as the one with the largest similarity measure (or smallest dissimilarity measure) between all weight vectors w_i and the input vector x. If the Euclidean distance is chosen as the dissimilarity measure, then the winning unit c satisfies the following equation:

$$\|x - w_c\| = \min_i \|x - w_i\|$$

where the index c refers to the winning unit.

Step2: Let N B_c denote a set of index corresponding to a neighborhood around winner c. The weights of the winner and its neighboring units are then updated by

$$\Delta wi = \eta(x - w_i), \quad i \in N\,B_c$$

where η is a small positive learning rate. Instead of defining the neighborhood of a winning unit, we can use a neighborhood function $\Omega c(i)$ around a winning unit c. For instance, the Gaussian function can be used as the neighborhood function:

$$\Omega c(i) = \exp\left(\frac{-\|p_i - p_c\|^2}{2\,\sigma^2}\right)$$

where p_i and p_c are the positions of the output units i and c, respectively, and σ reflects the scope of the neighborhood. By using the neighborhood function, the update formula can be rewritten as

$$\Delta wi = \eta\Omega_c(i)(x - w_i),$$

where i is the index for all output units.

The most well-known application of Kohonen self-organizing networks is Kohonen's attempt to construct a neural phonetic typewriter (Kohonen, 1988) that is capable of transcribing speech into written text from an unlimited vocabulary, with an accuracy of 92% to 97%. The network has also been used to learn ballistic arm movements (Ritter & Schulten, 1987).

We first illustrate the use of self-organizing networks with a very simple example. We have as initial data a set of 100 uniformly distributed points in the range $[0, \pi/2]$, as shown in Figure 5.7. After training a one-dimensional Kohonen self-organizing map of 10 neurons for 1000 epochs, we arrived to the final configuration shown in Figure 5.8. In this figure we can see the final distribution of the neurons, which represent the centers of the clusters. From this last figure we can notice that the points were organized around 10 evenly distributed clusters.

Our second example illustrates the application of a bi-dimensional Kohonen self-organizing map. We have as initial training data a set of 1000 randomly distributed points in the square $[-1, 1] \times [-1, 1]$, this initial situation is shown in Figure 5.9. After applying a 5×6 bi-dimensional self-organizing map for 1000 epochs, we arrived to the final configuration shown in Figure 5.10. In this last figure we can see very clearly the distribution of the 30 centers of the clusters after training. In this case, the final configuration of the clusters is not uniformly distributed due to the characteristics of the initial distribution of points.

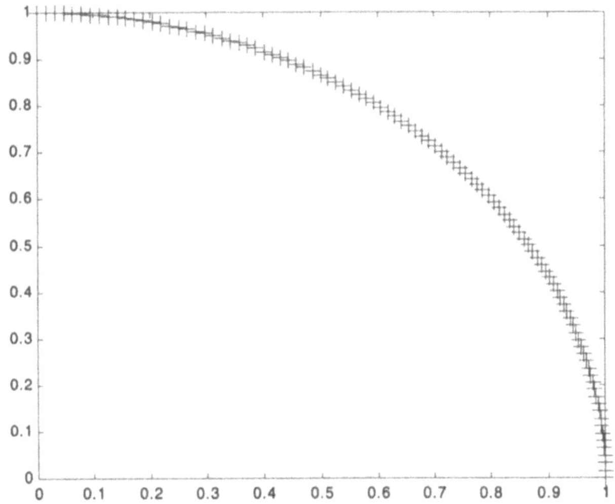

Figure 5.7 Initial data set of 100 uniformly distributed points in the range [0, π/2].

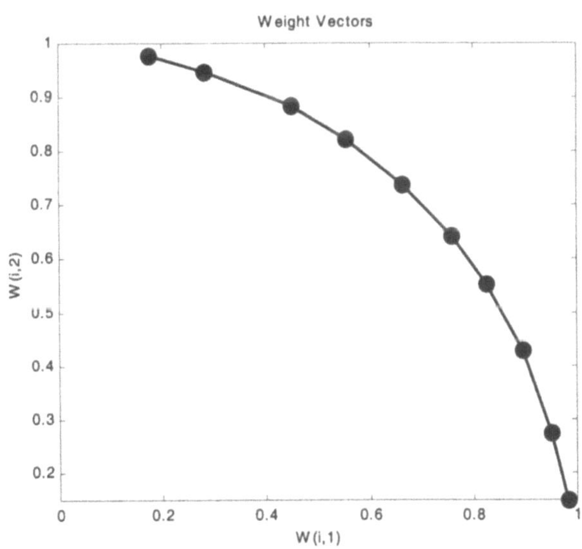

Figure 5.8 After training a self-organizing map of 10 neurons for 1000 epochs.

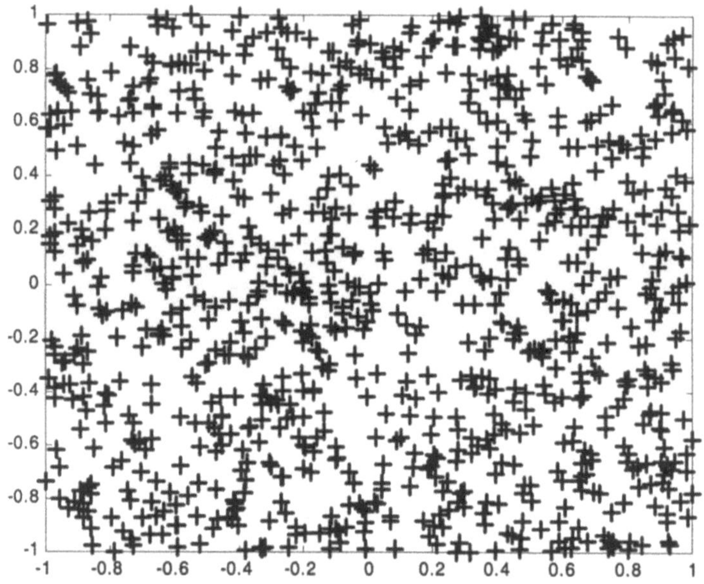

Figure 5.9 Initial set of 1000 distributed points in the square [-1, 1] × [-1, 1]

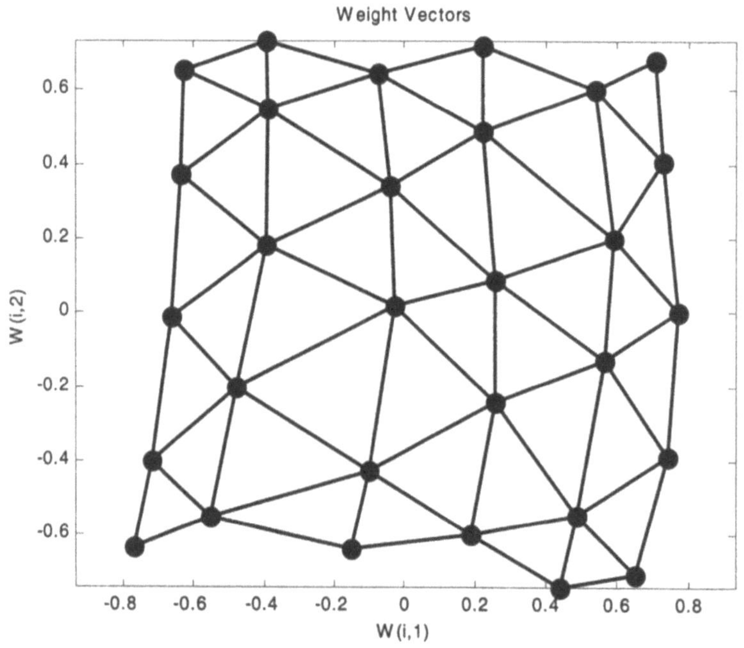

Figure 5.10 Final configuration of the clusters after training the map.

5.3 Learning Vector Quantization

Learning Vector Quantization (LVQ) is an adaptive data classification method based on training data with desired class information (Kohonen, 1989). Although a supervised training method, LVQ employs unsupervised data-clustering techniques (e.g., competitive learning, introduced in Section 5.1) to preprocess the data set and obtain cluster centers.

LVQ's network architecture closely resembles that of a competitive learning network, except that each output unit is associated with a class. Figure 5.11 presents an example, where the input dimension is 2 and the input space is divided into four clusters. Two of the clusters belong to class 1, and the other two clusters belong to class 2. The LVQ learning algorithm involves two steps. In the first step, an unsupervised learning data clustering method is used lo locate several cluster centers without using the class information. In the second step, the class information is used to fine-tune the cluster centers to minimize the number of misclassified cases.

During the first step of unsupervised learning, any of the data clustering techniques introduced in this chapter can be used to identify the cluster centers (or weight vectors leading to output units) to represent the data set with no class information. The numbers of clusters can either be specified a priori or determined via a cluster technique capable of adaptively adding new clusters when necessary. Once the clusters are obtained, their classes must be labeled before moving to the second step of supervised learning. Such labeling is achieved by the so-called "voting method" (i.e., a cluster is labeled class k if it has data points belonging to class k as a majority within the cluster). The clustering process for LVQ is based on the general assumption that similar input patterns generally belong to the same class.

During the second step of supervised learning, the cluster centers are fine-tuned to approximate the desired decision surface. The learning method is straightforward. First, the weight vector (or cluster center) **w** that is closest to the input vector **x** must be found. If **x** and **w** belong to the same class, we move **w** toward **x**; otherwise we move **w** away from the input vector **x**.

After learning, an LVQ network classifies an input vector by assigning it to the same class as the output unit that has the weight vector (cluster center) closest to the input vector. Figure 5.12 illustrates a possible distribution of data set and weights after training.

A sequential description of the LVQ method is the following:

Step 1: Initialize the cluster centers by a clustering method.

Step 2: Label each cluster by the voting method.

Step 3: Randomly select a training input vector **x** and find k such that $\| \mathbf{x} - \mathbf{w}_k \|$ is a minimum.

Step 4: If **x** and $\mathbf{w}_k$ belong to the same class, update $\mathbf{w}_k$ by

$$\Delta \mathbf{w}_k = \eta \, (x - \mathbf{w}_k).$$

Otherwise, update $\mathbf{w}_k$ by

$$\Delta \mathbf{w}_k = - \eta \, (x - \mathbf{w}_k).$$

The learning rate η is a positive small constant and should decrease with each iteration.

Step 5: If the maximum number of iterations is reached, stop. Otherwise, return to step 3.

Two improved versions of LVQ are available; both of them attempt to use the training data more efficiently by updating the winner and the runner-up (the next closest vector) under a certain condition. The improved versions are called LVQ2 and LVQ3 (Kohonen, 1990).

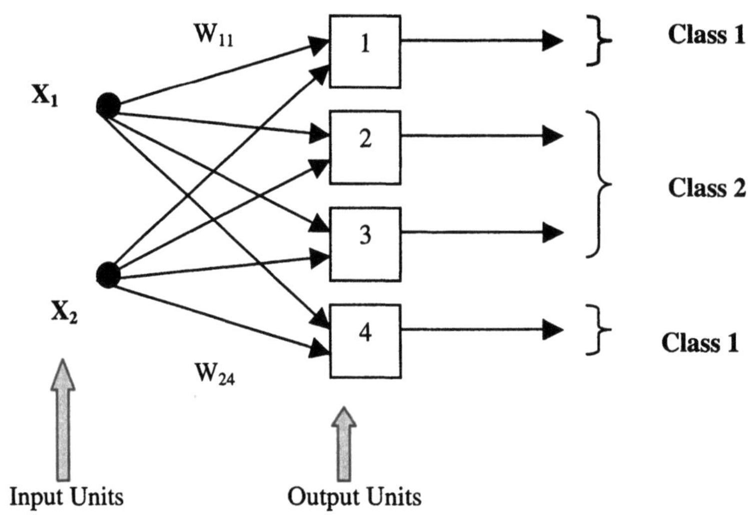

Figure 5.11 LVQ network representation.

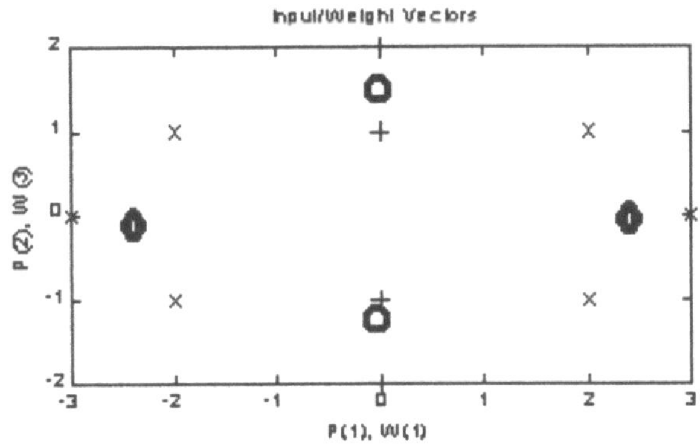

Figure 5.12 Data distribution and classification obtained by LVQ.

We will consider a simple example to illustrate the application of the LVQ network for classification purposes. Lets assume we have the data shown in Figure 5.13, which represents the 5 letters: O, C, A, L and I. We will apply a 5 neuron LVQ network to classify these letters. We show in Figure 5.14 the initial step of the LVQ training in which there is only one cluster at the center. In Figure 5.15 we show the final result after training the LVQ network. In this figure it is clearly seen that we have now five clusters, one for each of the letters. Also, the LVQ network has classified the letters in two types, which are the vowels (A and I) and the consonants (O, C and L). The vowels are indicated with circles and the consonants with diamonds.

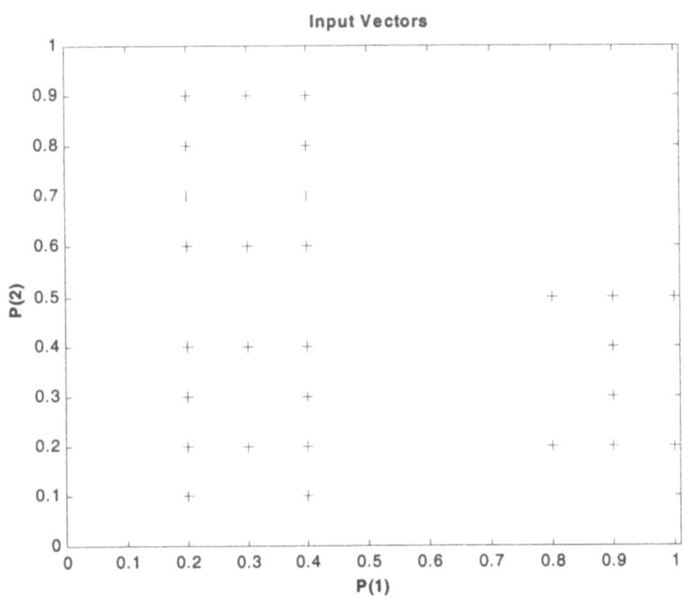

Figure 5.13 Data set representing 5 letters from the alphabet: O, C, A, L, and I.

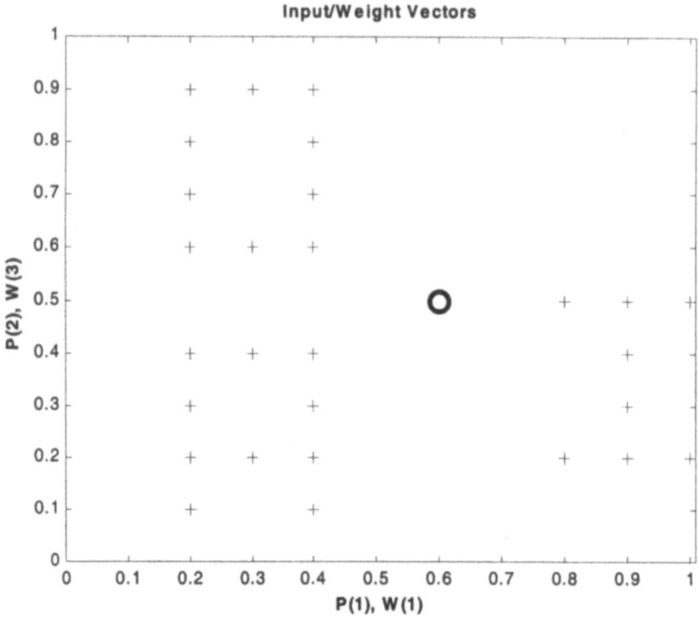

Figure 5.14 Initial approximation of the LVQ network with only one cluster.

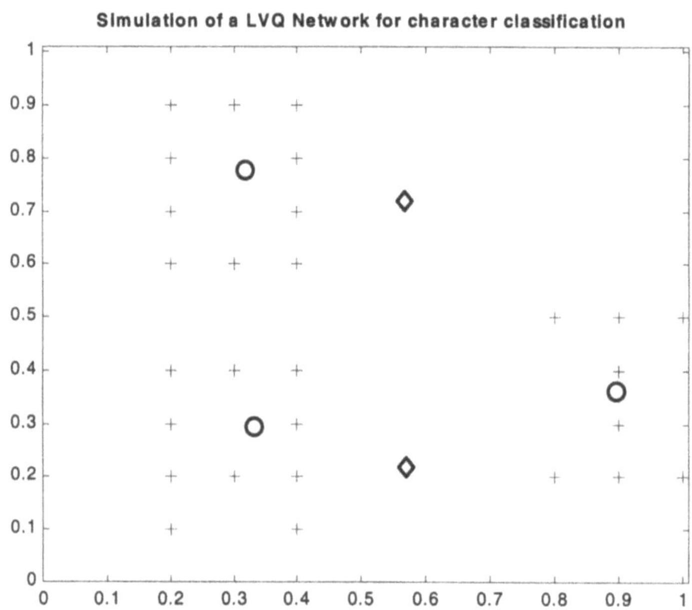

Figure 5.15 Final classification achieved with the LVQ network for the letters.

5.4 The Hopfield Network

In 1982, Hopfield proposed the so-called "Hopfield network", which possesses *auto-associative* properties. It is a recurrent (or fully interconnected) network in which all neurons are connected to each other, with the exception that no neuron has any connection to itself. In the network configuration, he embodied the *physical principle*, and set up an "energy function". The concept derives from a physical system (Hopfield, 1982):

"Any physical system whose dynamics in phase space is dominated by a substantial number of locally stable states to which it is attracted can therefore be regarded as a general content-addressable memory. The physical system will be a potentially useful memory if, in addition, any prescribed set of states can readily be made the stable states of the system."

The Hopfield neural network highlights a content-addressable memory and a tool for solving an optimization problem. These features of the Hopfield neural network will be described next.

The Hopfield net normally develops a number of locally stable points in state space. Because the dynamics of the network minimize "energy", other points in state space drain into the stable points (called "attractors" or "wells"), which are (possibly local) energy minima.

The Hopfield net realizes the operation of a "content-addressable" (auto-associative) memory in the sense that newly presented input patterns (or arbitrary initial states) can be connected to the appropriate patterns stored in memories (i.e., attractors or stable states). The presented input pattern vector cannot escape from a region, what we call a "basin of attraction", configured by each attractor (Figure 5.13). In other words, the neural network produces a desired memory pattern in response to a given pattern. The initial network state is assumed to lie within the reach of the fixed attractors' basins of attraction. That is, the entire configuration space is divided into different basins of attraction. Notably, an attractor is regarded as a fixed point if it is unique in state space. In general, however, an attractor may be "chaotic' (called a strange attractor) or may be a "limit cycle" of a periodic sequence of states. In terms of storage capacity, the number of memories is estimated to be nearly 10% to 20 % of the number of neurons in the Hopfield net (Denker, 1986).

We will illustrate the basic ideas of Hopfield networks with two simple examples. Lets consider first the case of a simple two-neuron Hopfield net that has two stable states. We show in Figure 5.14 the representation in state space of the two neurons, i.e., the attractors or wells of the dynamical system. We also show in Figure 5. 15 the simulation of the two-neuron Hopfield neural network, which clearly shows the two basins of attraction. We can interpret this figure as follows: initial points, which correspond to one of the basins of attraction, are pulled to that neuron (the arrows indicate the directions).

The example illustrates the very simple case of a two-neuron Hopfield network. Of course, increasingly complicated networks with more than two neurons may develop many stable states (attractors) to memorize several patterns.

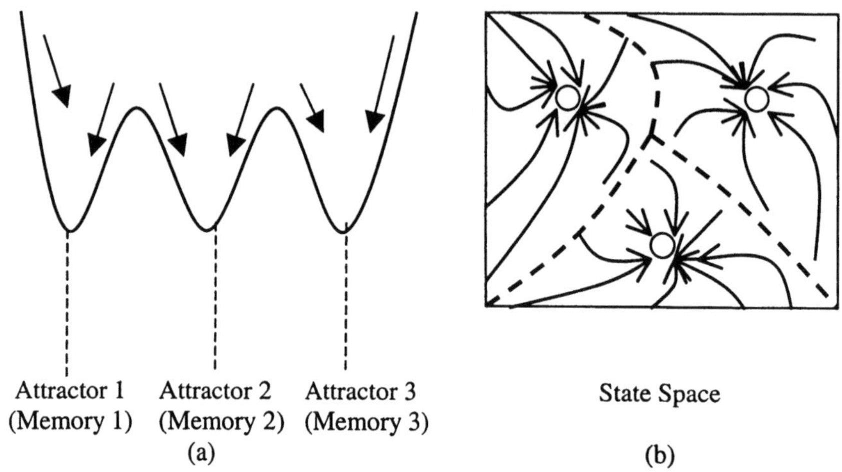

Figure 5.16 Images of energy terrains configured by three attractors in a Hopfield network. The circles denote stable states where patterns are memorized. The network state moves in the direction of the arrows to one of the attractors, which is determined by the starting point (i.e., the given input pattern).

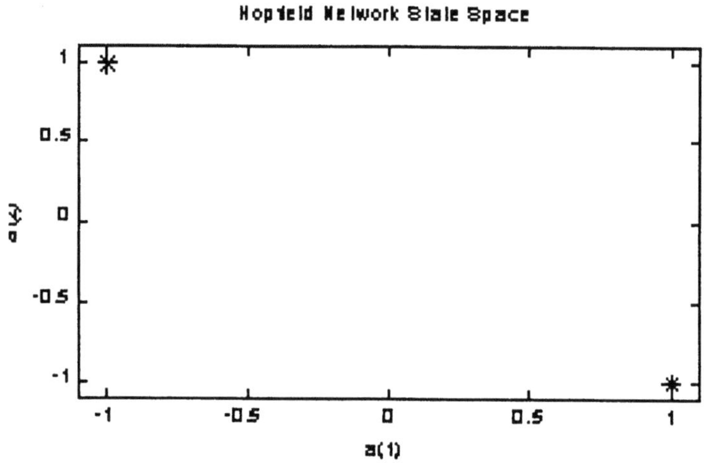

Figure 5.17 Hopfield network state space.

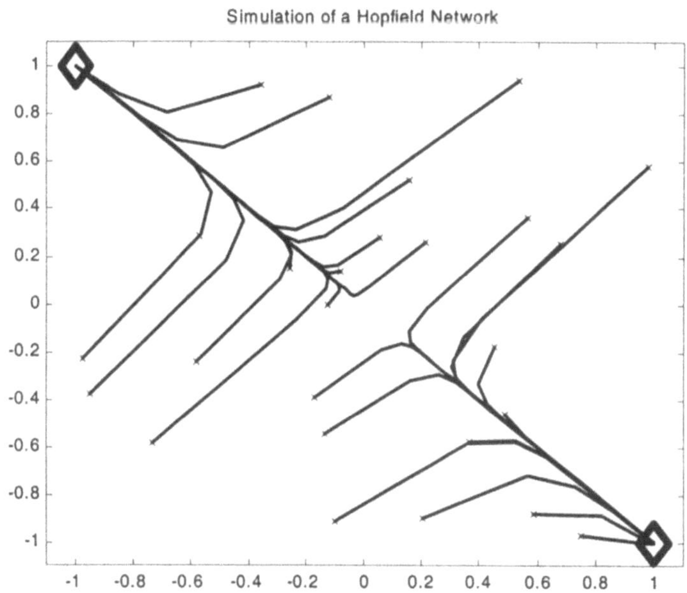

Figure 5.18 Simulation of a two-neuron Hopfield Network.

Lets consider as a second example three-neuron Hopfield network in three-dimensional space. The network has two stable states: the points (-1, 1, -1) and (1, 1, -1), which will divide the state space in two equal basins of attraction. We show in Figure 5.16 the state space for this Hopfield network. In Figure 5.17 we show how the initial points are attracted to one of these points. The directions are indicated with arrows. Actually, there is another stable point, which is (0, 1, -1), but the set of points, forming its basin of attraction, has measure zero.

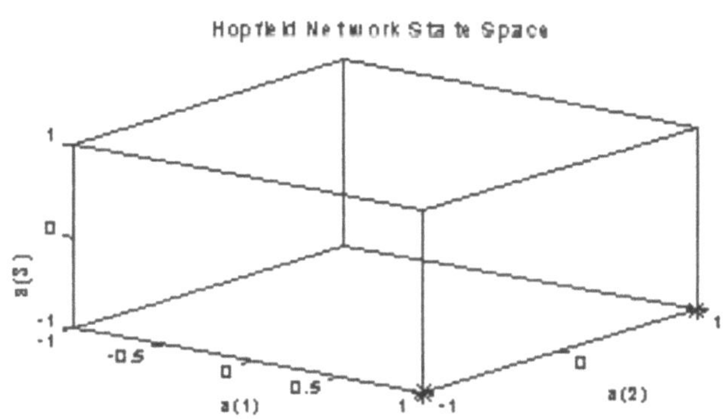

Figure 5.19 Three-dimensional state space for the Hopfield network.

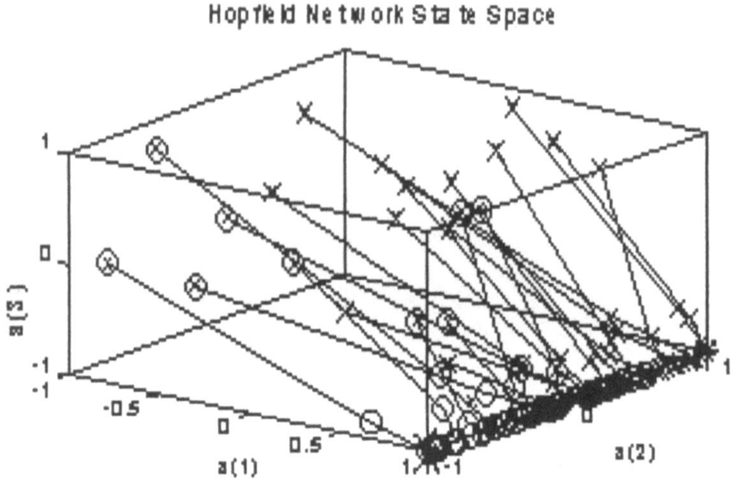

Figure 5.20 Simulation of the Hopfield network for different initial conditions.

5.5 Summary

We have presented in this chapter the basic concepts and theory of unsupervised learning neural networks. We have described in some detail competitive learning networks, Kohonen self-organizing networks, learning vector quantization networks, and Hopfield neural networks. Unsupervised learning is useful for analyzing data without having desired outputs; the neural networks evolve to capture density characteristics of a data set. We have shown some examples of how to use this type of unsupervised networks. We will describe in future chapters some applications of unsupervised neural networks to real world problems related to intelligent manufacturing.

Chapter 6

Genetic Algorithms and Simulated Annealing

This chapter introduces the basic concepts and notation of genetic algorithms and simulated annealing, which are two basic search methodologies that can be used for modelling and simulation of complex non-linear dynamical systems. Since both techniques can be considered as general purpose optimization methodologies, we can use them to find the mathematical model which minimizes the fitting errors for a specific problem. On the other hand, we can also use any of these techniques for simulation if we exploit their efficient search capabilities to find the appropriate parameter values for a specific mathematical model. We also describe in this chapter the application of genetic algorithms to the problem of finding the best neural network or fuzzy system for a particular problem. We can use a genetic algorithm to optimize the weights or the architecture of a neural network for a particular application. Alternatively, we can use a genetic algorithm to optimize the number of rules or the membership functions of a fuzzy system for a specific problem. These are two important application of genetic algorithms, which will be used in later chapters to design intelligent intelligent systems for controlling real world dynamical systems.

Genetic algorithms and simulated annealing have been used extensively for both continuous and discrete optimization problems (Jang, Sun & Mizutani, 1997). Common characteristics shared by these methods are described next.

- Derivative freeness: These methods do not need functional derivative information to search for a set of parameters that minimize (or maximize) a given objective function. Instead they rely exclusively on repeated evaluations of the objective function, and the subsequent search direction after each evaluation follows certain heuristic guidelines.

- Heuristic guidelines: The guidelines followed by these search procedures are usually based on simple intuitive concepts. Some of these concepts are motivated by so-called nature's wisdom, such as evolution and thermodynamics.

- Flexibility: Derivative freeness also relieves the requirement for differentiable objective functions, so we can use as complex an objective function as a specific application might need, without sacrificing too much in extra coding and computation time. In some cases, an objective function can even include the structure of a data-fitting model itself, which may be a fuzzy model.

- Randomness: These methods are stochastic, which means that they use random number generators in determining subsequent search directions. This element of randomness usually gives rise to the optimistic view that these methods are "global optimizers" that will find a global optimum given enough computing time. In theory, their random nature does make the probability of finding an optimal solution nonzero over a fixed amount of computation time. In practice, however, it might take a considerable amount of computation time.

- Analytic opacity: It is difficult to do analytic studies of these methods, in part because of their randomness and problem-specific nature. Therefore, most of our knowledge about them is based on empirical studies.

- Iterative nature: These techniques are iterative in nature and we need certain stopping criteria to determine when to terminate the optimization process. Let K denote an iteration count and f_k denote the best objective function obtained at count k; common stopping criteria for a maximization problem include the following:

    1) Computation time: a designated amount of computation time, or number of function evaluations and/or iteration counts is reached.
    2) Optimization goal: f_k is less than a certain preset goal value.
    3) Minimal improvement: $f_k - f_{k-1}$ is less than a preset value.
    4) Minimal relative improvement: $(f_k - f_{k-1})/ f_{k-1}$ is less than a preset value.

Both genetic algorithms (GAs) and simulated annealing (SA) have been receiving increasing amounts of attention due to their versatile optimization capabilities for both continuous and discrete optimization problems. Moreover, both of them are motivated by so-called "nature's wisdom": GAs are based on the concepts of natural selection and evolution; while SA originated in annealing processes found in thermodynamics and metallurgy.

6.1 Genetic Algorithms

Genetic algorithms (GAs) are derivative-free optimization methods based on the concepts of natural selection and evolutionary processes (Goldberg, 1989). They were first proposed and investigated by John Holland at the University of Michigan (Holland, 1975). As a general-purpose optimization tool, GAs are moving out of academia and finding significant applications in many areas. Their popularity can be attributed to their freedom from dependence on functional derivatives and their incorporation of the following characteristics:

- GAs are parallel-search procedures that can be implemented on parallel processing machines for massively speeding up their operations.
- GAs are applicable to both continuous and discrete (combinatorial) optimization problems.
- GAs are stochastic and less likely to get trapped in local minima, which inevitably are present in any optimization application.
- GAs' flexibility facilitates both structure and parameter identification in complex models such as fuzzy inference systems or neural networks.

GAs encode each point in a parameter (or solution) space into a binary bit string called a "chromosome", and each point is associated with a "fitness value" that, for maximization, is usually equal to the objective function evaluated at the point. Instead of a single point, GAs usually keep a set of points as a "population", which is then evolved repeatedly toward a better overall fitness value. In each generation, the GA constructs a new population using "genetic operators" such as crossover and mutation; members with higher fitness values are more likely to survive and to participate in mating (crossover) operations. After a number of generations, the population contains members with better fitness values; this is analogous to Darwinian models of evolution by random mutation and natural selection. GAs and their variants are sometimes referred to as methods of "population-based optimization" that improve performance by upgrading entire populations rather than individual members. Major components of GAs include encoding schemes, fitness evaluations, parent selection, crossover operators, and mutation operators; these are explained next.

Encoding schemes: These transform points in parameter space into bit string representations. For instance, a point (11, 4, 8) in a three-dimensional parameter space can be represented as a concatenated binary string:

$$\underbrace{1011}_{11}\quad\underbrace{0100}_{4}\quad\underbrace{1000}_{8}$$

in which each coordinate value is encoded as a "gene" composed of four binary bits using binary coding. other encoding schemes, such as gray coding, can also be used and, when necessary, arrangements can be made for encoding negative, floating-point, or discrete-valued numbers. Encoding schemes provide a way of translating problem-specific knowledge directly into the GA framework, and thus play a key role in determining GAs' performance. Moreover, genetic operators, such as crossover and mutation, can and should be designed along with the encoding scheme used for a specific application.

Fitness evaluation: The first step after creating a generation is to calculate the fitness value of each member in the population. For a maximization problem, the fitness value f_i of the ith member is usually the objective function evaluated at this member (or point). We usually need fitness values that are positive, so some kind of monotonical scaling and/or translation may by necessary if the objective function is not strictly positive. Another approach is to use the rankings of members in a population as their fitness values. The advantage of this is that the objective function does not need to be accurate, as long as it can provide the correct ranking information.

Selection: After evaluation, we have to create a new population from the current generation. The selection operation determines which parents participate in producing offspring for the next generation, and it is analogous to "survival of the fittest" in natural selection. Usually members are selected for mating with a selection probability proportional to their fitness values. The most common way to implement this is to set the selection probability equal to:

$$f_i / \sum_{k=1}^{k=n} f_k,$$

where n is the population size. The effect of this selection method is to allow members with above-average fitness values to reproduce and replace members with below-average fitness values.

Crossover: To exploit the potential of the current population, we use "crossover" operators to generate new chromosomes that we hope will retain good features from the previous generation. Crossover is usually applied to selected pairs of parents with a probability equal to a given "crossover rate". "One-point crossover" is the most basic crossover operator, where a crossover point on the genetic code is selected at random and two parent chromosomes are interchanged at this point. In "two-point crossover", two crossover points are selected and the part of the chromosome string between these two points is then swapped to generate two children. We can define n-point crossover similarly. In general, (n-1)-point crossover is a special case of n-point crossover. Examples of one-and two-point crossover are shown in Figure 6.1.

crossover point

100	11110	⇒	100	10010
101	10010		101	11110

(a)

1	0011	110	⇒	1	0110	110
1	0110	010		1	0011	010

(b)

Figure 6.1 Crossover operators: (a) one-point crossover; (b) two-point crossover.

Mutation: Crossover exploits current gene potentials, but if the population does not contain all the encoded information needed to solve a particular problem, no amount of gene mixing can produce a satisfactory solution. For this reason, a "mutation" operator capable of spontaneously generating new chromosomes is included. The most common way of implementing mutation is to flip a bit with a probability equal to a very low given "mutation rate". A mutation operator can prevent any single bit from converging to a value throughout the entire population and, more important, it can prevent the population from converging and stagnating at any local optima. The mutation rate is usually kept low so good chromosomes obtained from crossover are not lost. If the mutation rate is high (above 0.1), GA performance will approach that of a primitive random search. Figure 6.2 provides an example of mutation.

Mutated bit
↓
10011110 ⇒ 10011010

Figure 6.2 Mutation operator.

In the natural evolutionary process, selection, crossover, and mutation all occur in the single act of generating offspring. Here we distinguish them clearly to facilitate implementation of and experimentation with GAs.

Based on the aforementioned concepts, a simple genetic algorithm for maximization problems is described next.

Step 1: Initialize a population with randomly generated individuals and evaluate the fitness value of each individual.

Step 2: Perform the following operations:
(a) Select two members from the population with probabilities proportional to their fitness values.
(b) Apply crossover with a probability equal to the crossover rate.

(c) Apply mutation with a probability equal to the mutation rate.

(d) Repeat (a) to (d) until enough members are generated to form the next generation.

Step 3: Repeat steps 2 and 3 until a stopping criterion is met.

Figure 6.3 is a schematic diagram illustrating how to produce the next generation from the current one.

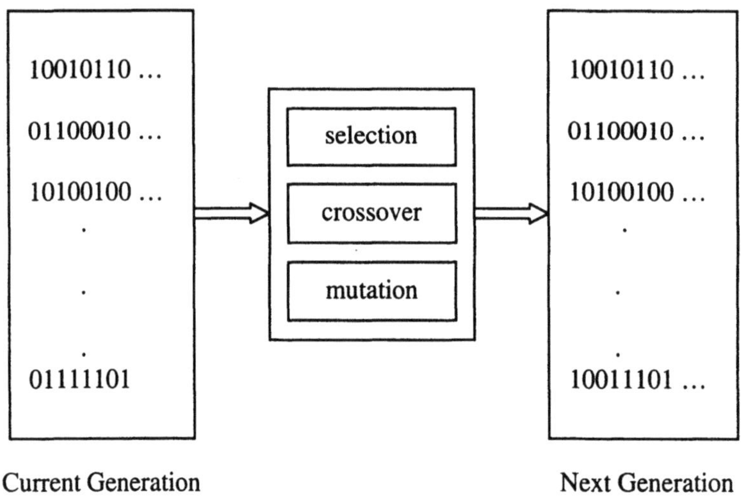

Figure 6.3 Producing the next generation in GAs.

Lets consider a simple example to illustrate the application of the basic genetic algorithm. Will consider the maximization of the "peaks" function, which is given by the following equation:

$$Z = f(x, y) = 3(1 - x)^2 e^{-x2-(y+1)2} - 10(x/5 - x^3 - y^5) e^{-x2-y2} - (1/3) e^{-(x+1)2 - y2} . \quad (6.1)$$

The surface plot of this function is shown in Figure 6.4. To use Gas to find the maximum of this function, we first confine the search domain to the square [-3, 3] × [-3, 3]. We use 8-bit binary coding for each variable. Each generation in our GA implementation contains 30 points or individuals. We use a simple one-point crossover scheme with crossover rate equal to 0.9. We choose uniform mutation with mutation rate equal to 0.05. Figure 6.5 shows a plot of the best, average, and poorest values of the objective function across 30 generations. Figure 6.6 shows the contour plot of the "peaks" function with the final population distribution after 30 generations. We can appreciate from these figures how the GA is able to find the maximum value of 8.

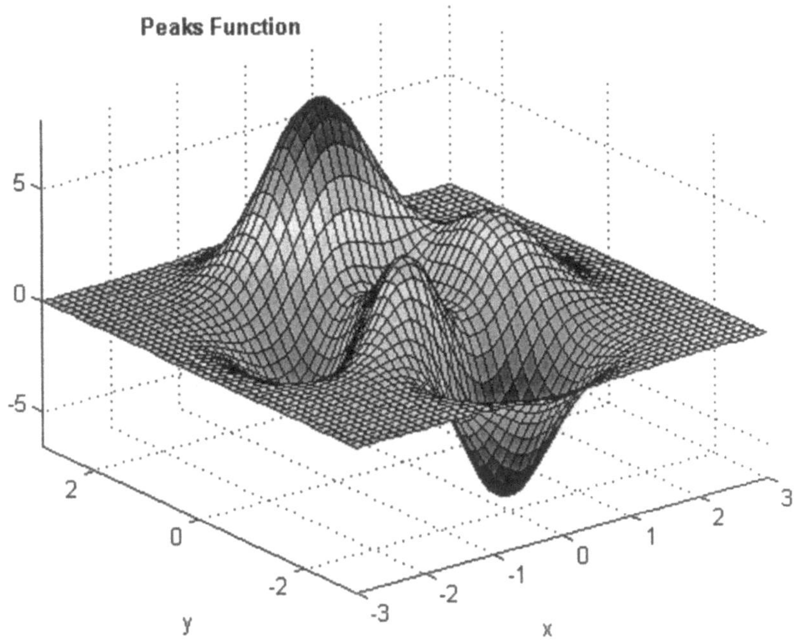

Figure 6.4 Surface plot of the "peaks" function.

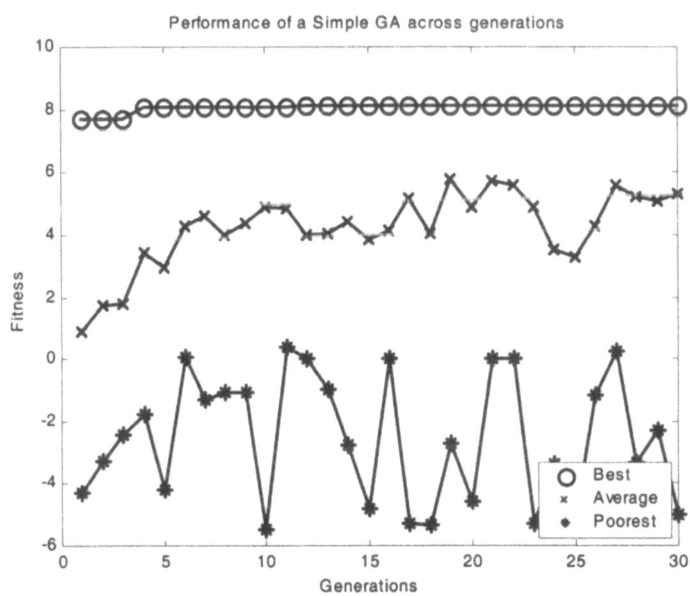

Figure 6.5 Performance of the genetic algorithm across generations.

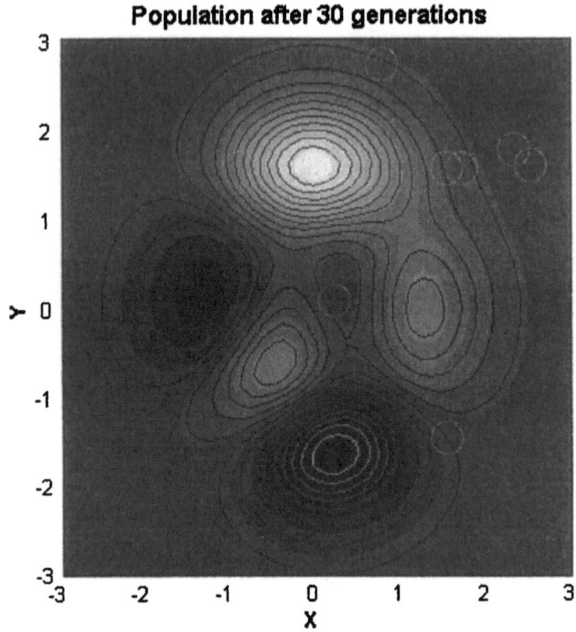

Figure 6.6 Contour plot of the "peaks" function with the final population.

Now lets consider a more complicated example. We will consider the Rosenbrock's valley, which is a classic optimization problem. The global optimum is inside a long, narrow, parabolic shaped flat valley. To find the valley is trivial, however convergence to the global optimum is difficult and hence this problem has been repeatedly used in assess the performance of optimization algorithms. In this case, the function is given by the following equation

$$f_2 = \Sigma^{n-1}{}_i\ 100\,(x_2 - x^2{}_1)^2 + (1 - x_1)^2 \qquad -2 \le x_i \le 2 \qquad (6.2)$$

global minimum: $x_i = 1$ $f(x) = 0$.

We show in Figure 6.7 the plot of Rosenbrock's valley function. We will apply a simple genetic algorithm with the same parameters as in the previous example, i.e. same population size, mutation rate, and crossover rate. The only parameter we will change is the maximum number of generations, which will now be of 250. Figure 6.8 shows the performance of the simple genetic algorithm for Rosenbrock's function. We can notice from this figure that the GA is able to converge in about 200 generations to a best value of 0.00034306. We have to mention that the computer program to obtain these results was implemented in the MATLAB programming language.

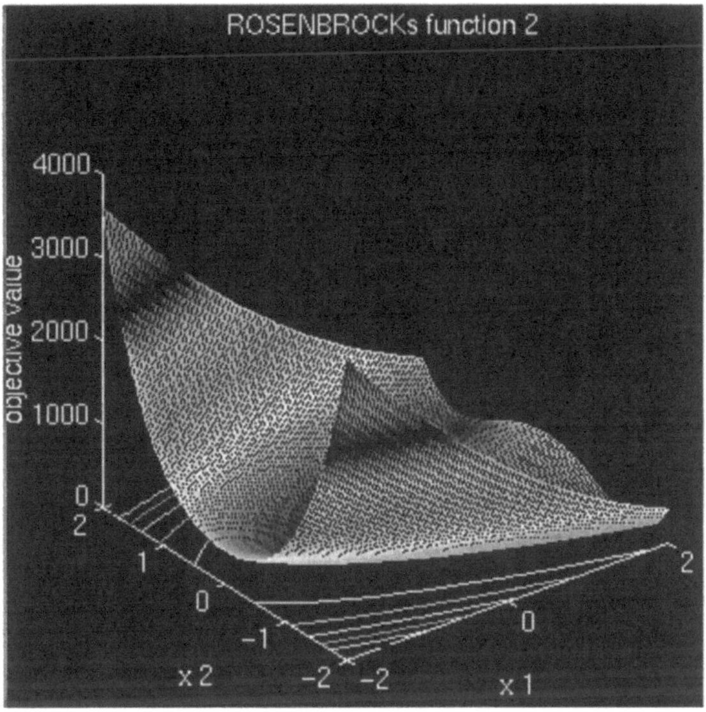

Figure 6.7 Plot of the Rosenbrock's valley.

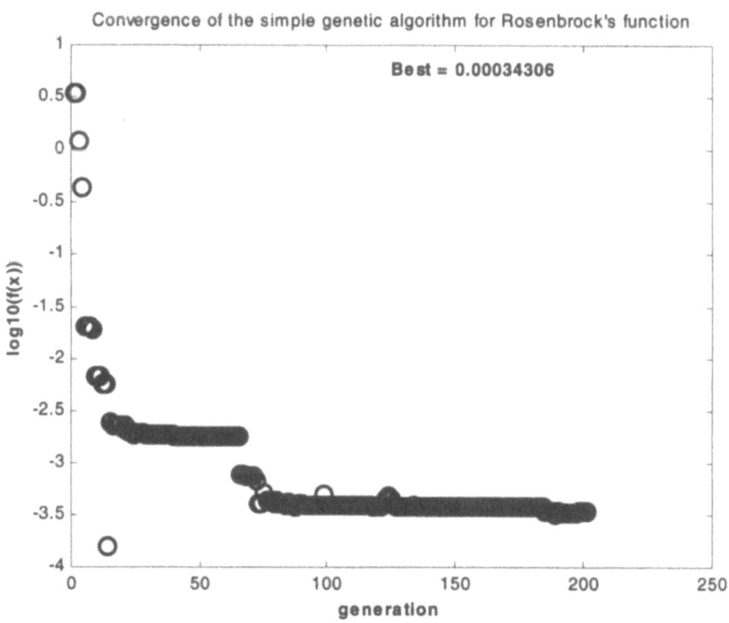

Figure 6.8 Performance of the genetic algorithm for Rosenbrock's valley.

6.2 Modifications to Genetic Algorithms

The GA mechanism is neither governed by the use of differential equations nor does it behave like a continuous function. However, it possesses the unique ability to search and optimize a solution for complex system, where other mathematical oriented techniques may have failed to compile the necessary design specifications. Due to its evolutionary characteristics, a standard GA may not be flexible enough for a practical application, and an engineering insight is always required whenever a GA is applied. This becomes more apparent where the problem to be tackled is complicated, multi-tasking and conflicting. Therefore, a means of modifying the GA structure is sought in order to meet the design requirements. There are many facets of operational modes that can be introduced.

6.2.1 Chromosome Representation

The problem to be tackled varies from one to the other. The coding of chromosome representation may vary according to the nature of the problem itself. In general, the bit string encoding (Holland, 1975) is the most classic method used by GA researchers because of its simplicity and traceability. The conventional GA operations and theory (scheme theory) are also developed on the basis of this fundamental structure. Hence, this representation is adopted in many applications. However, one minor modification can be suggested in that a Gray code may be used instead of the binary coding. Hollstien (1971) investigated the use of GA for optimizing functions of two variables based on a Gray code representation, and discovered that this works slightly better than the normal binary representation.

More recently, a direct manipulation of real-value chromosomes (Janikow & Michalewicz, 1991 and Wright, 1991) raised considerable interest. This representation was introduced especially to deal with real parameter problems. The work currently taking place by Janikow and Michalewicz indicates that the floating point representation would be faster in computation and more consistent from the basis of run-to-run. At the same time, its performance can be enhanced by special operators to achieve high accuracy (Michalewicz, 1996).

6.2.2 Objective Functions and Fitness

An objective function is a measuring mechanism that is used to evaluate the status of a chromosome. This is a very important link to relate the GA and the system concerned. Since each chromosome is individually going through the same calculations, the range of this value varies from one chromosome to another. To maintain uniformity, the objective value O is mapped into a fitness value with a map Ψ where the domain of F is usually greater than zero.

$$\Psi : \mathrm{O} \rightarrow \mathrm{F} \tag{6.3}$$

Linear Scaling

The fitness value fi of chromosome i has a linear relationship with the objective value o_i as

$$f_i = ao_i + b \qquad (6.4)$$

where a and b are chosen to enforce the equality of the average objective value and the average fitness value, and cause maximum scaled fitness to be a specified multiple of the average fitness.

This method can reduce the effect of genetic drift by producing an extraordinarily good chromosome. However, it may introduce a negative fitness value, which must be avoided. Hence, the choice of a and b are dependent on the knowledge of the range of the objective values.

Sigma Truncation

This method avoids the negative fitness value and incorporates the problem dependent information into the scaling mechanism. The fitness value f_i of chromosome i is calculated according to

$$f_i = o_i - (\bar{o} - c\sigma) \qquad (6.5)$$

where c is small integer, $\bar{o}$ is the mean of the objective values, σ is the standard deviation in the population.

To prevent negative values of f, any negative result $f < 0$ is arbitrarily set to zero. Chromosomes whose fitness values are less than c (a small integer from the range 1 and 5) standard deviations from the average fitness value are not selected.

Power Law Scaling

The actual fitness value is taken as a specific power of the objective value, o^k_i

$$Fi = o^k_i \qquad (6.6)$$

Where k is in general problem dependent or even varying during the run (Guillies, 1985).

Ranking

There are other methods that can be used such as the Ranking scheme (Baker, 1987). The fitness values do not directly relate to their corresponding objective values, but to the ranks of the objective values.

Using this approach can help the avoidance of premature convergence and speed up the search when the population approaches convergence. On the other hand, it requires additional overheads in the in the GA computation for sorting chromosomes according to their objective values.

6.2.3 Selection Methods

To generate good offspring, a good parent selection mechanism is necessary. This is a process used for determining the number of trials for one particular individual used in reproduction. The chance of selecting one chromosome as a parent should be directly proportional to the number of offspring produced.

Baker (1987) presented three measures of performance of the selection methods: Bias, Spread, and Efficiency. *Bias* defines the absolute difference between individuals in actual and expected probability for selection. Optimal zero bias is achieved when an individual's probability equals its expected number of trials. *Spread* is a range in the possible number of trials that an individual may achieve. If g(i) is the actual number of trials due to each individual i, then the "minimum spread" is the smallest spread that theoretically permits zero bias, i.e.

$$G(i) \in \{ \lfloor et(i) \rfloor, \lceil et(i) \rceil \} \tag{6.7}$$

where et(i) is the expected number of trials of individual i, $\lfloor et(i) \rfloor$ is the floor and $\lceil et(i) \rceil$ is the ceiling. Thus the spread of a selection method measures its consistency. *Efficiency* is related to the overall time complexity of the algorithms.

The selection method should thus be achieving a zero bias whilst maintaining a minimum spread and not contributing to an increased time complexity of the GA.

Many selection techniques employ the "roulette wheel mechanism". The basic roulette wheel selection method is a stochastic sampling with replacement (SSR) technique. The segment size and selection probability remain the same throughout the selection phase and the individuals are selected according to the above procedures. SSR tends to give zero bias but potentially inclines to a spread that is unlimited.

Stochastic Sampling with Partial Replacement (SSPR) extends upon SSR by resizing a chromosome's segment if it is selected. Each time a chromosome is selected, the size of its segment is reduced by a certain factor. If the segment size becomes negative, then it is set to zero. This provides an upper bound on the spread of $\lceil et(i) \rceil$ but with a zero lower bound and a higher bias. The roulette wheel selection methods can generally be implemented with a time complexity of the order of *NlogN* where *N* is the population size.

Stochastic Universal Sampling (SUS) is another single-phase sampling method with minimum spread, zero bias and the time complexity is in the order of *N* (Baker, 1987). SUS uses an N equally spaced pointer, where N is the number of selections required. The population is shuffled randomly and a single random number in the range $[0, F_{sum}/N]$ is generated, ptr, where F_{sum} is the sum of the individuals' fitness values. An individual is thus guaranteed to be selected a minimum of $\lfloor et(i) \rfloor$ times and no more than $\lceil et(i) \rceil$, thus achieving minimum spread. In addition, as individuals are selected entirely based on their position in the population, SUS has zero bias.

6.2.4 Genetic Operations

Crossover

Although the one-point crossover method was inspired by biological processes, it has one major drawback in that certain combinations of schema cannot be combined in some situations (Michalewicz, 1996).

A multi-point crossover can be introduced to overcome this problem. As a result, the performance of generating offspring is greatly improved. Another approach is the uniform crossover. This generates offspring from the parents, based on a randomly generated crossover mask. The resulting offspring contain a mixture of genes from each parent. The number of effective crossing points is not fixed, but will be averaged at L/2 (where L is the chromosome length).

The preference for using which crossover techniques is still arguable. However, De Jong (1975) concluded that a two-point crossover seemed to be an optimal number for multi-point crossover. Since then, this has been contradicted by Spears and De Jong (1991) as a two-point crossover could perform poorly if the population has largely being converged because of any reduced crossover productivity. A general comment was that each of these crossover operators was particularly useful for some classes of problems and quite poor for others, and that the one-point crossover was considered a "loser" experimentally.

Crossover operations can be directly adopted into the chromosome with real number representation. The only difference would be if the string were composed of a series of real numbers instead of binary numbers.

Mutation

Originally, mutation was designed only for the binary-represented chromosome. To adopt the concept of introducing variations into the chromosome, a random mutation (Michalewicz, 1996) has been designed for a real number chromosome:

$$g = g + \psi(\mu, \sigma) \tag{6.8}$$

where g is the real number gene; ψ is a random function, which may be Gaussian or normally distributed; μ is the mean and σ is the variance of the random function.

Operational rates settings

The choice of an optimal probability operation rate for crossover and mutation is another controversial debate for both analytical and empirical investigations. The increase of crossover probability would cause the recombination of building blocks to rise, and at the same time, it also increases the disruption of good chromosomes. On the other hand, should the mutation probability increase, this would transform the genetic search into a random search, but would help to reintroduce the lost genetic material.

6.2.5 Parallel Genetic Algorithm

Considering that the GA already possesses an intrinsic parallelism architecture, there is no extra effort to construct a parallel computational framework. Rather, the GA can be fully exploited in its parallel structure to gain the required speed for practical applications.

There are a number of GA-based parallel methods to enhance the computational speed (Cantu Paz, 1995). The methods of parallelization can be classified as Global, Migration and Diffusion. These categories reflect different ways in which parallelism can be exploited in the GA as well as the nature of the population structure and recombination mechanisms used.

<u>Global GA</u>
Global GA treats the entire population as a single breeding mechanism. This can be implemented on a shared memory multiprocessor or distributed memory computer. On a shared memory multiprocessor, chromosomes are stored in the shared memory. Each processor accesses the particular assigned chromosome and returns the fitness values without any conflicts. It should be noted that there is some synchronization needed between generation to generation. It is necessary to balance the computational load among the processors using a dynamic scheduling algorithm.

On a distributed memory computer, the population can be stored in one processor to simplify the genetic operators. This is based on the farmer-worker architecture. The farmer processor is responsible for sending chromosomes to the worker processors for the purpose of fitness evaluation. It also collects the results from them, and applies the genetic operators for producing the next generation.

<u>Migration GA</u>
This is another parallel processing method for computing the GA. The migration GA divides the population into a number of sub-populations, each of which is treated as a separate breeding unit under the control of a conventional GA. To encourage the proliferation of good genetic material throughout the whole population, migration between the sub-populations occurs from time to time. The required parameters for successful migration are the "migration rate" and the "migration interval". The migration rate governs the number of individuals to be migrated. The migration interval affects the frequency of migrations. The values of these parameters are intuitively chosen rather than based on some rigorous scientific analysis. In general, the occurrence of migration is usually set at a predetermined constant interval that is governed by migration intervals. We illustrate the concept of migration in Figure 6.9.

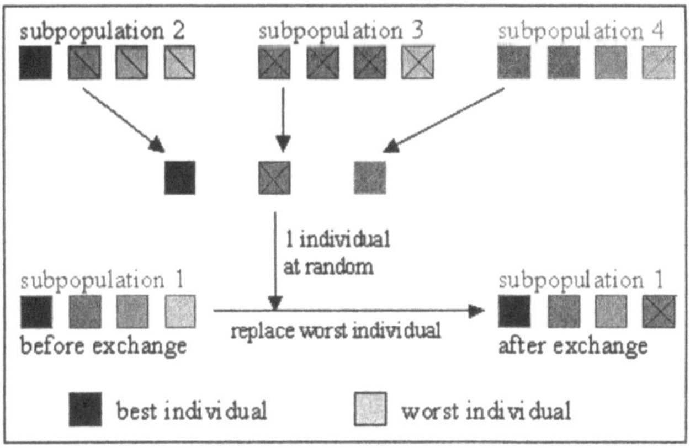

Figure 6.9 Detailed description of the migration concept.

The migration GA is well suited to parallel implementation on Multiple Instruction Multiple Data (MIMD) machines. The architecture of hypercubes and rings are commonly used for this purpose. Given the range of possible population topologies and migration paths between them, efficient communication networks should thus be possible on most parallel architectures. This applies to small multiprocessor platforms or even the clustering of networked workstations.

Diffusion GA

The Diffusion GA considers the population as a single continuous structure. Each individual is assigned to a geographic location on the population surface and usually placed in a two-dimensional grid. This is because of the topology of the processing element in many massively parallel computers that are constructed in this form. The individuals are allowed to breed with individuals contained in a small local neighborhood. This neighborhood is usually chosen from immediately adjacent individuals on the population surface and is motivated by the practical communication restrictions of parallel computers.

We will illustrate the ideas of parallel genetic algorithms with the well-known harvest optimization problem. We will apply a migration GA to solve this optimization problem. The harvest system is a one-dimensional equation of growth with one constraint:

$$x(k+1) = a\,x(k) - u(k) \qquad k = 1,\dots, N, \tag{6.9}$$

such that $x(0) = x(N)$.

The objective function for minimization is therefore defined as:

$$F(u) = -\sum_{k=1}^{N} [u(k)]^{1/2}. \tag{6.10}$$

We will apply a multi-population genetic algorithm with migration and real-valued representation of the individuals. We used 20 decision variables, a crossover rate of 1.0 and a mutation rate of 1/number of variables. We used 8 subpopulations with 20 individuals each, and a migration rate of 0.2. The maximum number of generations was specified at 200. We show in Figure 6.10 the initial population of individuals. In Figure 6.11 we show the simulation results after 200 generations of the GA.

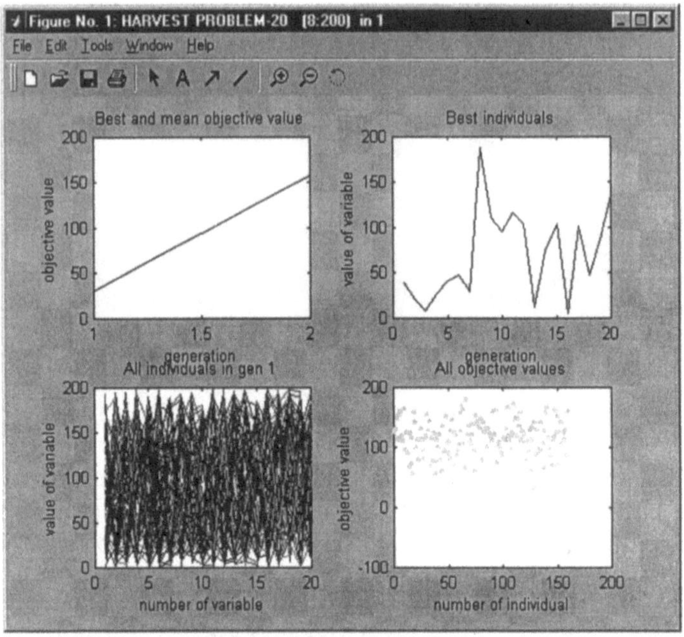

Figure 6.10 Initial population and subpopulations for the migration GA.

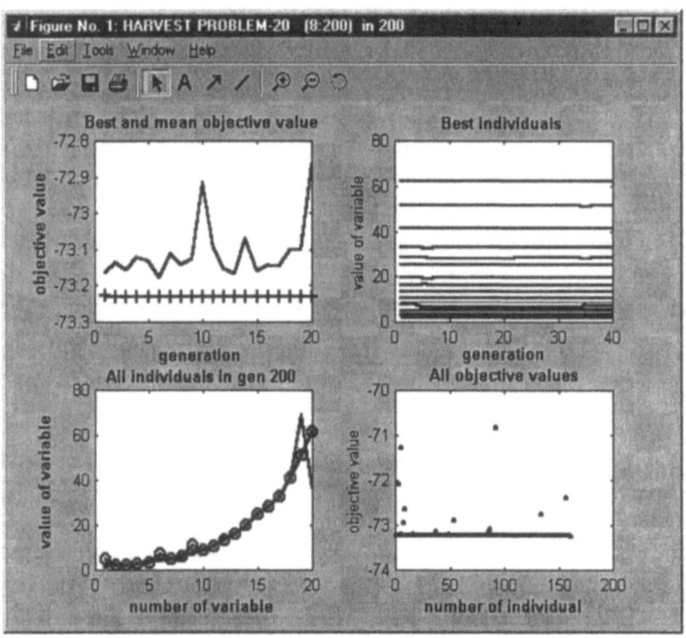

Figure 6.11 Final population for the migration GA after 200 generations.

6.3 Simulated Annealing

"Simulated Annealing" (SA) is another derivative-free optimization method that has recently drawn much attention for being as suitable for continuous as for discrete (combinatorial) optimization problems (Otten & Ginneken, 1989). When SA was first proposed (Kirkpatrick, Gelatt & Vecchi, 1983) it was mostly known for its effectiveness in finding near optimal solutions for large-scale combinatorial optimization problems, such as traveling salesperson problems and placement problems. Recent applications of SA and its variants (Ingber & Rosen, 1992) also demonstrate that this class of optimization approaches can be considered competitive with other approaches when there are continuous optimization problems to be solved.

Simulated annealing was derived from physical characteristics of spin glasses (Kirkpatrick, Gelatt & Vecchi, 1983). The principle behind simulated annealing is analogous to what happens when metals are cooled at a controlled rate. The slowly falling temperature allows the atoms in the molten metal to line themselves up and form a regular crystalline structure that has high density and low energy. But if the temperature goes down too quickly, the atoms do not have time to orient themselves into a regular structure and the result is a more amorphous material with higher energy.

In simulated annealing, the value of an objective function that we want to minimize is analogous to the energy in a thermodynamic system. At high temperatures, SA allows function evaluations at faraway points and it is likely to accept a new point with higher energy. This corresponds to the situation in which high-mobility atoms are trying to orient themselves with other nonlocal atoms and the energy state can occasionally go up. At low temperatures, SA evaluates the objective function only at local points and the likelihood of it accepting a new point with higher energy is much lower. This is analogous to the situation in which the low-mobility atoms can only orient themselves with local atoms and the energy state is not likely to go up again.

Obviously, the most important part of SA is the so-called "annealing schedule" or "cooling schedule", which specifies how rapidly the temperature is lowered from high to low values. This is usually application specific and requires some experimentation by trial-and-error.

Before giving a detailed description of SA, first we shall explain the fundamental terminology of SA.

Objective function: An objective function f(.) maps an input vector x into a scalar E:

$$E = f(x),$$

where each x is viewed as a point in an input space. The task of SA is to sample the input space effectively to find a x that minimizes E.

Generating function: A generating function g(. , .) specifies the probability density function of the difference between the current point and the next point to be visited. Specifically, Δx (= x_{new} - x) is a random variable with probability density function $g(\Delta x,T)$, where T is the temperature. For common SA (especially in combinatorial optimization applications), g(. , .) is usually a function independent of the temperature T.

Acceptance function: After a new point x_{new} has been evaluated, SA decides whether to accept or reject it based on the value of an acceptance function h(. , .). The most frequently used acceptance function is the "Boltzmann probability distribution":

$$h(\Delta E, T) = \frac{1}{1 + \exp(\Delta E / (cT))} \tag{6.11}$$

where c is a system-dependent constant, T is the temperature, and ΔE is the energy difference between x_{new} and x:

$$\Delta E = f(x_{new}) - f(x)$$

The common practice is to accept x_{new} with probability $h(\Delta E , T)$.

Annealing schedule: An annealing schedule regulates how rapidly the temperature T goes from high to low values, as a function of time or iteration counts. The exact interpretation of "high" and "low" and the specification of a good annealing schedule require certain problem-specific physical insights and/or trial-and-error. The easiest way of setting an annealing schedule is to decrease the temperature T by a certain percentage at each iteration.

The basic algorithm of simulated annealing is the following:

Step 1: Choose a start point x and set a high starting temperature T. Set the iteration count k to 1.

Step 2: Evaluate the objective function $E = f(x)$.

Step 3: Select Δx with probability determined by the generating function $g(\Delta x, T)$. Set the new point x_{new} equal to $x + \Delta x$.

Step 4: Calculate the new value of the objective function: $E_{new} = f(x_{new})$.

Step 5: Set x to x_{new} and E to E_{new} with probability determined by the acceptance function $h(\Delta E , T)$, where $\Delta E = E_{new} - E$.

Step 6: Reduce the temperature T according to the annealing schedule (usually by simply setting T equal to ηT, where η is a constant between 0 and 1).

Step 7: Increment iteration count k. If k reaches the maximum iteration count, stop the iterating. Otherwise, go back to step 3.

In conventional SA, also known as "Boltzmann machines", the generating function is a Gaussian probability density function:

$$g(\Delta x, T) = (2\pi T)^{-n/2} \exp[-\| \Delta x \|^2 / (2T)] \quad (6.12)$$

where Δx ($= x_{new} - x$) is the deviation of the new point from the current one, T is the temperature, and n is the dimension of the space under exploration. It has been proven (Geman & Geman, 1984) that a Boltzman machine using the aforementioned generating function g(. , .) can find a global optimum of f(x) if the temperature T is reduced no faster than $T_0 / \ln k$.

Variants of Boltzmann machines include the "Cauchy machine" or "fast simulated annealing" (Szu & Hartley, 1987), where the generating function is the Cauchy distribution:

$$g(\Delta x) = \frac{T}{(\| \Delta x \|^2 + T^2)^{(n+1)/2}} \quad (6.13)$$

The fatter tail of the Cauchy distribution allows it to explore farther from the current point during the search process.

Another variant of the original SA, the so-called "very fast" simulated annealing (Ingber & Rosen, 1992), was designed for optimization problems in a constrained search space. Very fast simulated annealing has been reported to be faster than genetic algorithms on several test problems by the same authors.

Lets give a simple example of the use of SA for the traveling sales person problem (TSP). We will consider finding the solution for 25-city TSP problem. The simulation results are shown in Figure 6.12.

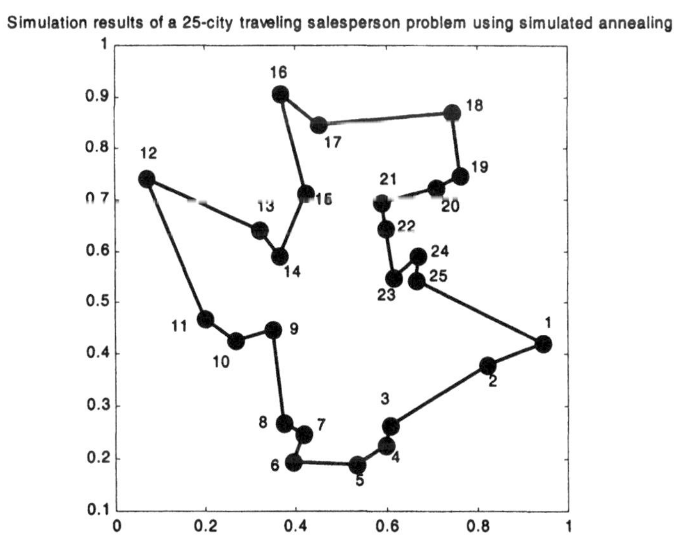

Figure 6.12 The result of a 25-city traveling sales person problem with SA.

6.4 Applications of Genetic Algorithms

The genetic algorithms described in the previous sections are very simple, but variations of these algorithms have been used in a large number of scientific and engineering problems and models (Mitchell, 1996). Some examples follow.

- Optimization: genetic algorithms have been used in a wide variety of optimization tasks, including numerical optimization and such combinatorial optimization problems as circuit layout and job-shop scheduling.
- Automatic Programming: genetic algorithms have been used to evolve computer programs for specific tasks, and to design other computational structures such as cellular automata and sorting networks.
- Machine Learning: genetic algorithms have been used for many machine learning applications, including classification and prediction tasks, such as the prediction of weather or protein structure. Genetic algorithms have also been used to evolve aspects of particular machine learning systems, such as weights for neural networks, rules for learning classifier systems or symbolic production systems, and sensors for robots.
- Economics: genetic algorithms have been used to model processes of innovation, the development of bidding strategies, and the emergence of economic markets.
- Immune Systems: genetic algorithms have been used to model various aspects of natural immune systems, including somatic mutation during an individual's lifetime and the discovery of multi-gene families during evolutionary time.
- Ecology: genetic algorithms have used to model ecological phenomena such as biological arms races, host-parasite co-evolution, symbiosis, and resource flow.
- Social Systems: genetic algorithms have been used to study evolutionary aspects of social systems, such as the evolution of social behavior in insect colonies, and, more generally, the evolution of cooperation and communication in multi-agent systems.

This list is by no means exhaustive, but it gives the flavor of the kinds of things genetic algorithms have been used for, both in problem solving and in scientific contexts. Because of their success in these and other areas, interest in genetic algorithms has been growing rapidly in the last several years among researchers in many disciplines.

We will describe bellow the application of genetic algorithms to the problem of evolving neural networks, which is a very important problem in designing the particular neural network for a problem.

6.4.1 Evolving Neural Networks

Neural Networks are biologically motivated approaches to machine learning, inspired by ideas from neuroscience. Recently, some efforts have been made to use genetic algorithms to evolve aspects of neural networks (Mitchell, 1996).

In its simplest feedforward form, a neural network is a collection of connected neurons in which the connections are weighted, usually with real-valued weights. The network is presented with an activation pattern on its input units, such as a set of numbers representing features of an image to be classified. Activation spreads in a forward direction from the input units through one or more layers of middle units to the output units over the weighted connections. This process is meant to roughly mimic the way activation spreads through networks of neurons in the brain. In a feedforward network, activation spreads only in a forward direction, from the input layer through the hidden layers to the output layer. Many people have also experimented with "recurrent" networks, in which there are feedback connections between layers.

In most applications, the neural network learns a correct mapping between input and output patterns via a learning algorithm. Typically the weights are initially set to small random values. Then a set of training inputs is presented sequentially to the network. In the backpropagation learning procedure, after each input has propagated through the network and an output has been produced, a "teacher" compares the activation value at each output unit with the correct values, and the weights in the network are adjusted in order to reduce the difference between the network's output and the correct output. This type of procedure is known as "supervised learning", since a teacher supervises the learning by providing correct output values to guide the learning process.

There are many ways to apply genetic algorithms to neural networks. Some aspects that can be evolved are the weights in a fixed network, the network architecture (i.e., the number of neurons and their interconnections can change), and the learning rule used by the network.

6.4.1.1 Evolving Weights in a Fixed Network

David Montana and Lawrence Davis (1989) took the first approach of evolving the weights in a fixed network. That is, Montana and Davis were using the genetic algorithm instead of backpropagation as a way of finding a good set of weights for a fixed set of connections. Several problems associated with the backpropagation algorithm (e.g., the tendency to get stuck at local minima, or the unavailability of a "teacher" to supervise learning in some tasks) often make it desirable to find alternative weight training schemes.

Montana and Davis were interested in using neural networks to classify underwater sonic "lofargrams" (similar to spectrograms) into two classes: "interesting" and "not interesting". The networks were to be trained from a database containing lofargrams and classifications made by experts as to whether or not a given lofargram is "interesting". Each network had four input units, representing four parameters used by an expert system that performed the same classification. Each network had one output unit and two layers of hidden units (the first with seven units and the second with ten units). The networks were fully connected feedforward networks. In total there were 108 weighted connections between units. In addition, there were 18 weighted connections between the non-input units and a "threshold unit" whose outgoing links implemented the thresholding for each of the non-input units, for a total of 126 weights to evolve.

The genetic algorithm was used as follows. Each chromosome was a list of 126 weights. Figure 6.13 shows (for a much smaller network) how the encoding was done: the weights were read off the network in a fixed order (from left to right and from top to bottom) and placed in a list. Notice that each "gene" in the chromosome is a real number rather than a bit. To calculate the fitness of a given chromosome, the weights in the chromosome were assigned to the links in the corresponding network, the network was run on the training set (here 236 examples from the database), and the sum of the squares of the errors was returned. Here, an "error" was the difference between the desired output value and the actual output value. Low error meant high fitness in this case.

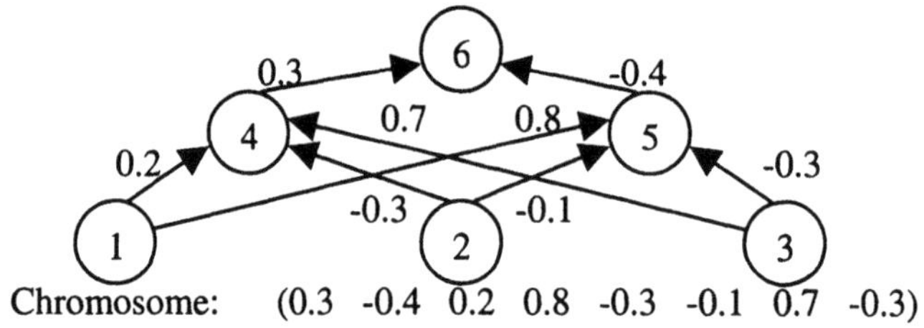

Figure 6.13 Encoding of network weights for the genetic algorithm.

An initial population of 50 weights vectors was chosen randomly, with each weight being between –1.0 and +1.0. Montana and Davis tried a number of different genetic operators in various experiments. The mutation and crossover operators they used for their comparison of the genetic algorithm with backpropagation are illustrated in Figures 6.14 and 6.15. The mutation operator selects n non-input units, and for each incoming link to those units, adds a random value between –1.0 and +1.0 to the weight on the link. The crossover operator takes two parent weight vectors, and for each non-input unit in the offspring vector, selects one of the parents at random and copies the weights on the incoming links from that parent to the offspring. Notice that only one offspring is created.

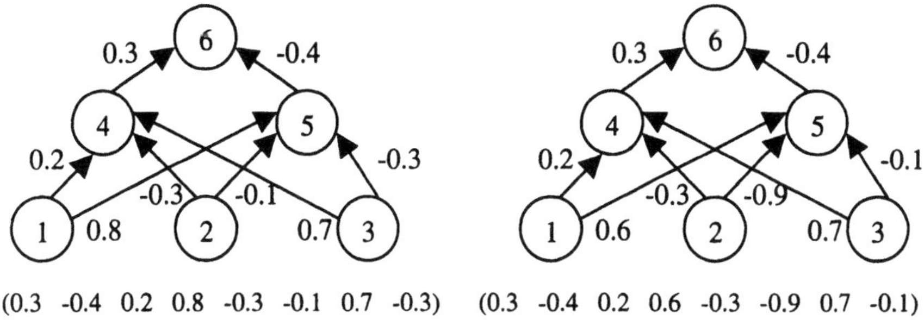

Figure 6.14 Illustration of the mutation method. The weights on incoming links to unit 5 are mutated.

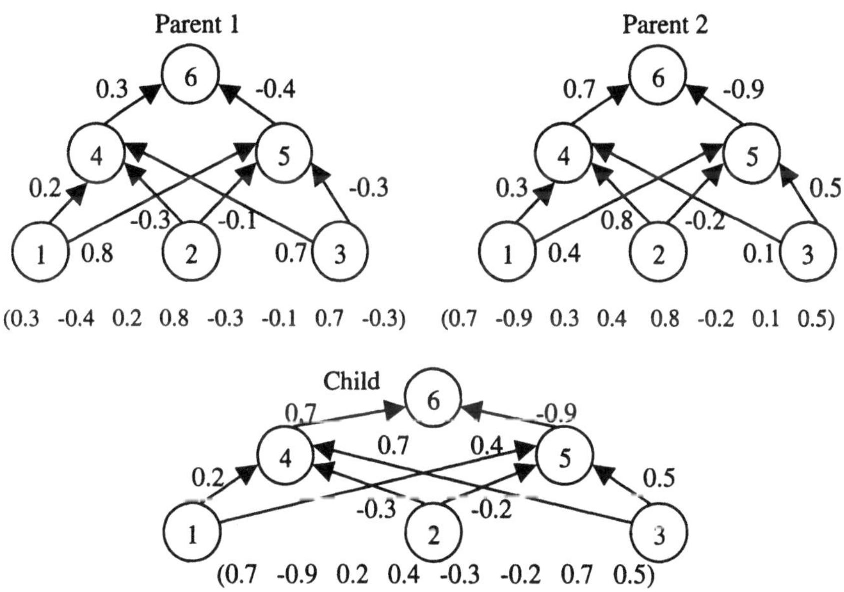

Figure 6.15 Illustration of the crossover method. In the child network shown here, the incoming links to unit 4 come from parent 1 and the incoming links 5 and 6 come from parent 2.

The performance of a genetic algorithm using these operators was compared with the performance of a backpropagation algorithm. The genetic algorithm had a population of 50 weight vectors, and a rank selection method was used. The genetic algorithm was allowed to run for 200 generations. The backpropagation algorithm was allowed to run for 5000 iterations, where one iteration is a complete epoch (a complete pass through the training data). Montana and Davis found that the genetic algorithm significantly outperforms backpropagation on this task, obtaining better weight vectors more quickly.

This experiment shows that in some situations the genetic algorithm is a better training method for neural networks than simple backpropagation. This does not mean that the genetic algorithm will outperform backpropagation in all cases. It is also possible that enhancements of backpropagation might help it overcome some of the problems that prevented it from performing as well as the genetic algorithm in this experiment.

6.4.1.2 Evolving Network Architectures

Neural network researchers know all too well that the particular architecture chosen can determine the success or failure of the application, so they would like very much to be able to automatically optimize the procedure of designing an architecture for a particular application. Many believe that genetic algorithms are well suited for this task (Mitchell, 1996). There have been several efforts along these lines, most of which fall into one of two categories: direct encoding and grammatical encoding. Under direct encoding a network architecture is directly encoded into a genetic algorithm chromosome. Under grammatical encoding, the genetic algorithm does not evolve network architectures; rather, it evolves grammars that can be used to develop network architectures.

Direct Encoding

The method of direct encoding is illustrated in work done by Geoffrey Miller, Peter Todd, and Shailesh Hedge (1989), who restricted their initial project to feedforward networks with a fixed number of units for which the genetic algorithm was used to evolve the connection topology. As is shown in Figure 6.16, the connection topology was represented by a NxN matrix (5x5 in Figure 6.16) in which each entry encodes the type of connection from the "from unit" to the "to unit". The entries in the connectivity matrix were either "0" (meaning no connection) or "L" (meaning a "learnable" connection). Figure 6.16 also shows how the connectivity matrix was transformed into a chromosome for the genetic algorithm ("0" corresponds to 0 and "L" to 1) and how the bit string was decoded into a network. Connections that were specified to be learnable were initialized with small random weights.

From unit		1	2	3	4	5
To unit	1	0	0	0	0	0
	2	0	0	0	0	0
	3	L	L	0	0	0
	4	L	L	0	0	0
	5	0	0	L	L	0

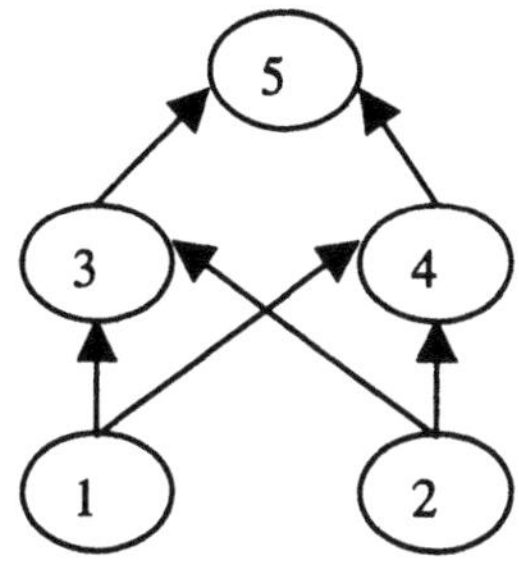

Chromosome: 0 0 0 0 0 0 0 0 0 0 1 1 0 0 0 1 1 0 0 0 0 0 1 1 0

Figure 6.16 Illustration of Miller, Todd, and Hedge's representation scheme.

Miller, Todd, and Hedge used a simple fitness-proportionate selection method and mutation (bits in the string were flipped with some low probability). Their crossover operator randomly chose a row index and swapped the corresponding rows between the two parents to create two offspring. The intuition behind that operator was similar to that behind Montana and Davis's crossover operator-each row represented all the incoming connections to a single unit, and this set was thought to be a functional building block of the network. The fitness of a chromosome was calculated in the same way as in Montana and Davis's project: for a given problem, the network was trained on a training set for a certain number of epochs, using backpropagation to modify the weights. The fitness of the chromosome was the sum of the squares of the errors on the training set at the last epoch. Again, low error translated to high fitness. Miller, Todd, and Hedge tried their genetic algorithm on several problems with very good results. The problems were relatively easy for multilayer neural networks to learn to solve under backpropagation. The networks had different number of units for different tasks; the goal was to see if the genetic algorithm could discover a good connection topology for each task. For each run the population size was 50, the crossover rate was 0.6, and the mutation rate was 0.005. In all cases, the genetic algorithm was easily able to find networks that readily learned to map inputs to outputs over the training set with little error. However, the tasks were too easy to be a rigorous test of this method-it remains to be seen if this method can scale up to more complex tasks that require much larger networks with many more interconnections.

Grammatical Encoding

The method of grammatical encoding can be illustrated by the work of Hiroaki Kitano (1990), who points out that direct encoding approaches become increasingly difficult to use as the size of the desired network increases. As the network's size grows, the size of the required chromosome increases quickly, which leads to problems both in performance and in efficiency. In addition, since direct encoding methods explicitly represent each connection in the network, repeated or nested structures cannot be represented efficiently, even though these are common for some problems.

The solution pursued by Kitano and others is to encode networks as grammars; the genetic algorithm evolves the grammars, but the fitness is tested only after a "development" step in which a network develops from the grammar. A grammar is a set of rules that can be applied to produce a set of structures (e.g., sentences in a natural language, programs in a computer language, neural network architectures).

Kitano applied this general idea to the development of neural networks using a type of grammar called a "graph-generation grammar", a simple example of which is given in Figure 6.17a. Here the right-hand side of each rule is a 2x2 matrix rather than a one-dimensional string. Each lower-case letter from a through p represents one of the 16 possible 2x2 arrays of ones and zeros. There is only one structure that can be formed from this grammar: the 8x8 matrix shown in Figure 6.17b. This matrix can be interpreted as a connection matrix for a neural network: a 1 in row i and column i means that unit i is present in the network and a 1 in row i and column , $i \neq j$, means that there is connection from unit i to unit j. The result is the network shown in Figure 6.17c which, with appropriate weights, computes the Boolean function XOR.

Kitano's goal was to have a genetic algorithm evolve such grammars. Figure 6.18 illustrates a chromosome encoding the grammar given in Figure 6.17a. The chromosome is divided up into separate rules, each of which consists of five elements. The first element is the left-hand side of the rule; the second through fifth elements are the four symbols in the matrix on the right-hand side of the rule. The possible values for each element are the symbols A-Z and a-p. The first element of the chromosome is fixed to be the start symbol, S; at least one rule taking S into a 2x2 matrix is necessary to get started in building a network from a grammar

The fitness of a grammar was calculated by constructing a network from the grammar, using backpropagation with a set of training inputs to train the resulting network to perform a simple task, and then, after training, measuring the sum of the squares of the errors made by the network on either the training set or a separate test set. The genetic algorithm used fitness-proportionate selection, multi-point crossover, and mutation. A mutation consisted of replacing one symbol in the chromosome with a randomly chosen symbol from the A-Z and a-p alphabets. Kitano used what he called "adaptive mutation": the probability of mutation of an offspring depended on the Hamming distance (number of mismatches) between the two parents. High distance resulted in low mutation, and vice versa. In this way, the genetic algorithm tended to respond to loss of diversity in the population by selectively raising the mutation rate.

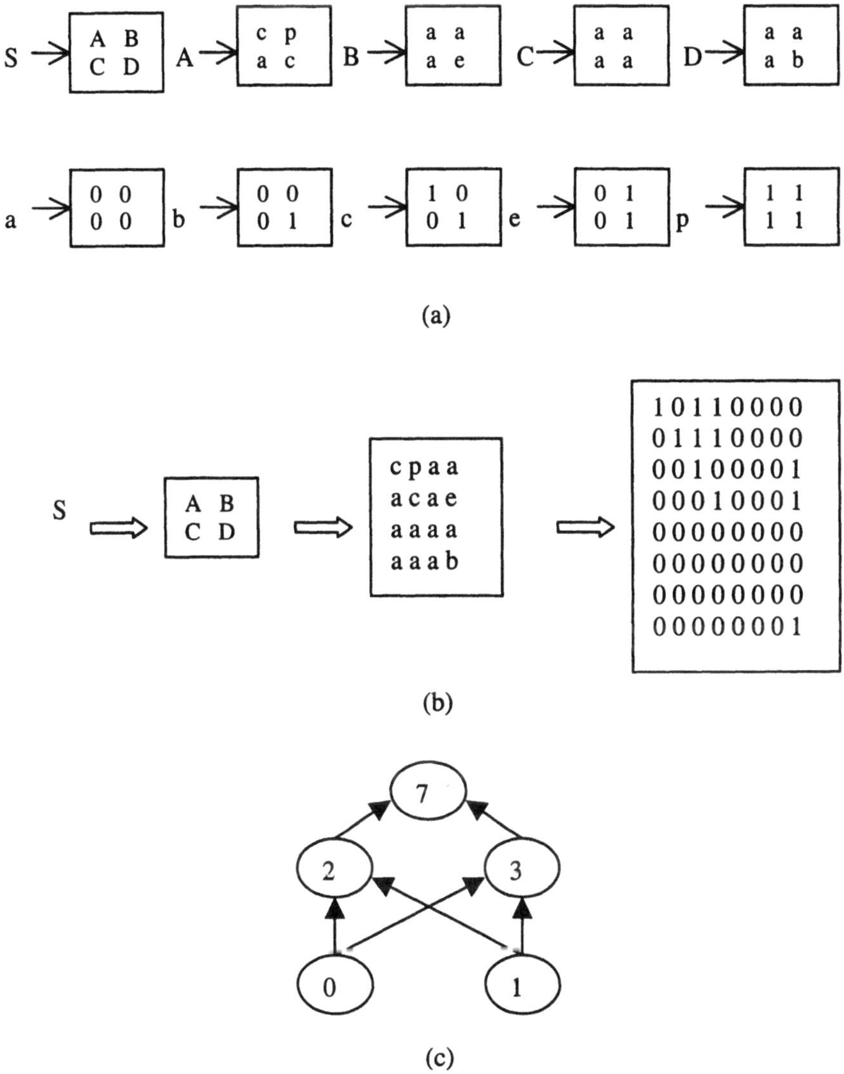

Figure 6.17 Illustration of Kitano's graph generation grammar for the XOR problem. (a) Grammatical rules. (b) A connection matrix is produced from the grammar. (c) The resulting network.

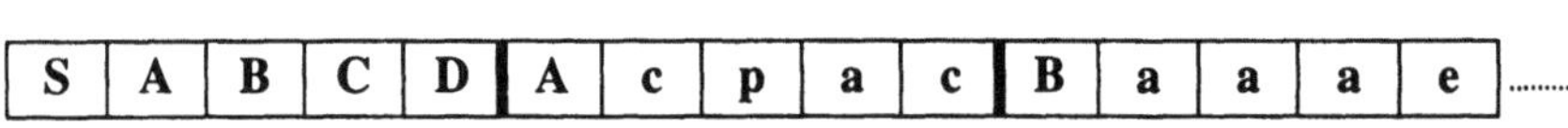

Figure 6.18 Illustration of a chromosome encoding a grammar.

Kitano (1990) performed a series of experiments on evolving networks for simple "encoder/decoder" problems to compare the grammatical and direct encoding approaches. He found that, on these relatively simple problems, the performance of a genetic algorithm using the grammatical encoding method consistently surpassed that of a genetic algorithm using the direct encoding method, both in the correctness of the resulting neural networks and in the speed with which they were found by the genetic algorithm. In the grammatical encoding runs, the genetic algorithm found networks with lower error rate, and found the best networks more quickly, than in direct encoding runs. Kitano also discovered that the performance of the genetic algorithm scaled much better with network size when grammatical encoding was used.

What accounts for the grammatical encoding method's apparent superiority? Kitano argues that the grammatical encoding method can easily create "regular", repeated patterns of connectivity, and that this is a result of the repeated patterns that naturally come from repeatedly applying grammatical rules. We would expect grammatical encoding approaches, to perform well on problems requiring this kind of regularity. Grammatical encoding also has the advantage of requiring shorter chromosomes, since the genetic algorithm works on the instructions for building the network (the grammar) rather that on the network structure itself. For complex networks, the later could be huge and intractable for any search algorithm.

Evolving a Learning Rule

David Chalmers (1990) took the idea of applying genetic algorithms to neural networks in a different direction: he used genetic algorithms to evolve a good learning rule for neural networks. Chalmers limited his initial study to fully connected feedforward networks with input and output layers only, no hidden layers. In general a learning rule is used during the training procedure for modifying network weights in response to the network's performance on the training data. At each training cycle, one training pair is given to the network, which then produces an output. At this point the learning rule is invoked to modify weights. A learning rule for a single layer, fully connected feedforward network might use the following local information for a given training cycle to modify the weight on the link from input unit i to output unit j:

a_i: the activation of input i
o_j: the activation of output unit j
t_j: the training signal on output unit j
w_{ij}: the current weight on the link from i to j
The change to make in the weight w_{ij} is a function of these values:
$\Delta w_{ij} = f(a_i, o_j, t_j, w_{ij})$.

The chromosomes in the genetic algorithm population encoded such functions.

Chalmers made the assumption that the learning rule should be a linear function of these variables and all their pairwise products. That is, the general form of the learning rule was

$$\Delta w_{ij} = k_0(k_1 w_{ij}+k_2 a_i+k_3 o_j+k_4 t_j+k_5 w_{ij} a_i+k_6 w_{ij} o_j+k_7 w_{ij} t_j+k_8 a_i o_j+k_9 a_i t_j+k_{10} o_j t_j).$$

The k_m ($1<m<10$) are constant coefficients, and k0 is a scale parameter that affects how much the weights can change on any cycle. Chalmer's assumption about the form of the learning rule came in part from the fact that a known good learning rule for such networks is the "delta rule". One goal of Chalmer's work was to see if the genetic algorithm could evolve a rule that performs as well as the delta rule.

The task of the genetic algorithm was to evolve values for the km's. The chromosome encoding for the set of km's is illustrated in Figure 6.19. The scale parameter k_0 is encoded as five bits, with the zeroth bit encoding the sign (1 encoding + and 0 encoding -) and the first through fourth bits encoding an integer n: $k_0 = 0$ if $n = 0$; otherwise $|k_0| = 2^{n-9}$. Thus k_0 can take on the values 0,+-1/256,+-1/128,...,+-32,+-64. The other coefficients km are encoded by three bits each, with the zeroth bit encoding the sign and the first and second bits encoding an integer n. For i=1,...,10, km=0 if n=0; otherwise $|k_m| = 2^{n-1}$.

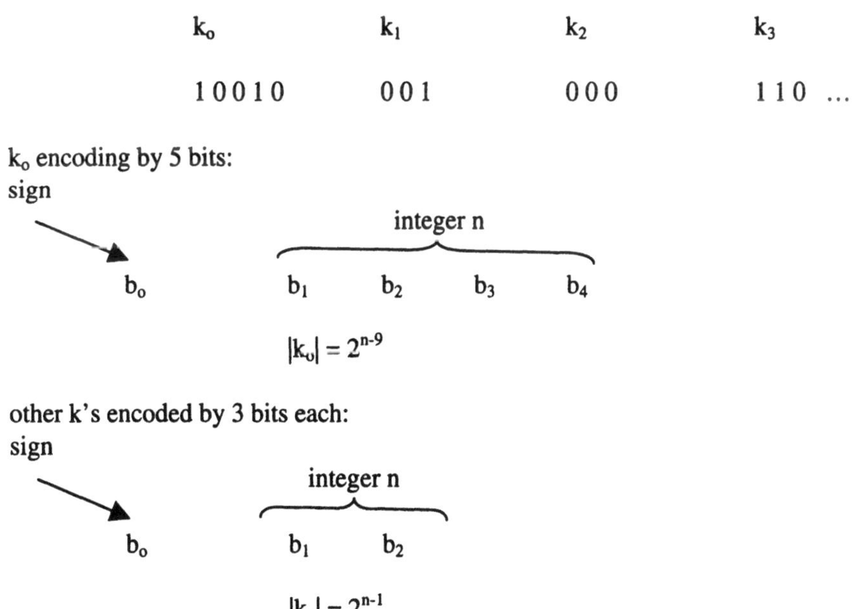

Figure 6.19 Illustration of the method for encoding the km's.

The fitness of each chromosome (learning rule) was determined as follows. A subset of 20 mappings was selected from the full set of 30 linear separable mappings. For each mapping, 12 training examples were selected. For each of these mappings, a network was created with the appropriate number of input units for the given mapping. The network's weights were initialized randomly. The network was run on the training set for some number of epochs (typically 10), using the learning rule specified by the chromosome. The performance of the learning rule on a given mapping was a function of the network's error on the training set, with low error meaning high performance. The overall fitness of the learning rule was a function of the average error of 20 networks over the chosen subset of 20 mappings. This fitness was then transformed to be a percentage, where a high percentage meant high fitness.

Using this fitness measure, the genetic algorithm was run on a population of 40 learning rules, with two-point crossover and standard mutation. The crossover rate was 0.8 and the mutation rate was 0.01. Typically, over 1000 generations, the fitness of the best learning rules in the population rose from between 40% and 60% in the initial generation to between 80% and 98%. The fitness of the delta rule is around 98%, and on one out of a total of ten run the genetic algorithm discovered this rule. On three of the ten runs, the genetic algorithm discovered slight variations of this rule with lower fitness.

These results show that, given a somewhat constrained representation, the genetic algorithm was able to evolve a successful learning rule for simple single layer networks. The extent to which this method can find learning rules for more complex networks remains an open question, but these results are a first step in that direction. Chalmers suggested that it is unlikely that evolutionary methods will discover learning methods that are more powerful than backpropagation, but he speculated that genetic algorithms might be a powerful method for discovering learning rules for unsupervised learning paradigms or for new classes of network architectures.

6.4.2 Evolving Fuzzy Systems

Ever since the very first introduction of the fundamental concept of fuzzy logic by Zadeh in 1973, its use in engineering disciplines has been widely studied. Its main attraction undoubtedly lies in the unique characteristics that fuzzy logic systems possess. They are capable of handling complex, non-linear dynamic systems using simple solutions. Very often, fuzzy systems provide a better performance than conventional non-fuzzy approaches with less development cost.

However, to obtain an optimal set of fuzzy membership functions and rules is not an easy task. It requires time, experience, and skills of the operator for the tedious fuzzy tuning exercise. In principle, there is no general rule or method for the fuzzy logic set-up. Recently, many researchers have considered a number of intelligent techniques for the task of tuning the fuzzy set.

Here, another innovative scheme is described (Man, Tang & Kwong, 1999). This approach has the ability to reach an optimal set of membership functions and rules without a known overall fuzzy set topology. The conceptual idea of this approach is to have an automatic and intelligent scheme to tune the membership functions and rules, in which the conventional closed loop fuzzy control strategy remains unchanged, as indicated in Figure 6.20.

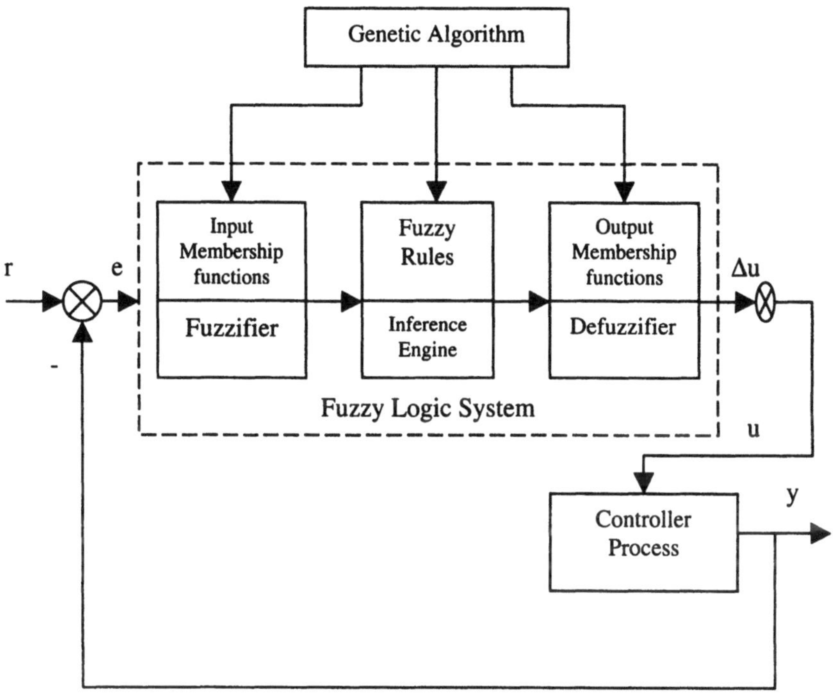

Figure 6.20 Genetic algorithm for a fuzzy control system.

In this case, the chromosome of a particular is shown in Figure 6.21. The chromosome consist of two types of genes, the control genes and parameter genes. The control genes, in the form of bits, determine the membership function activation, whereas the parameter genes are in the form of real numbers to represent the membership functions.

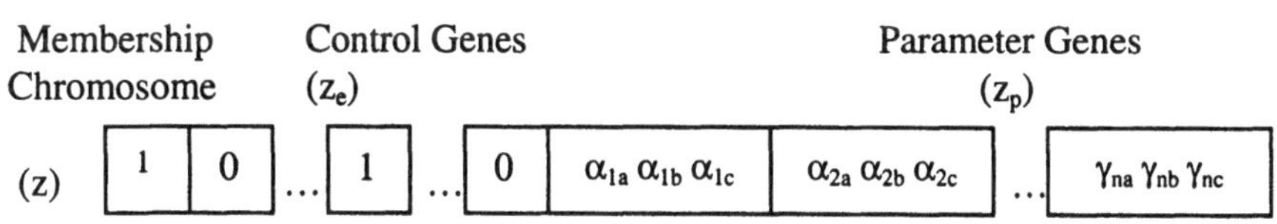

Figure 6.21 Chromosome structure for the fuzzy system.

To obtain a complete design for the fuzzy control system, an appropriate set of fuzzy rules is required to ensure system performance. At this point it should be stressed that the introduction of the control genes is done to govern the number of fuzzy subsets in the system.

Once the formulation of the chromosome has been set for the fuzzy membership functions and rules, the genetic operation cycle can be performed. This cycle of operation for the fuzzy control system optimization using a genetic algorithm is illustrated in Figure 6.22.

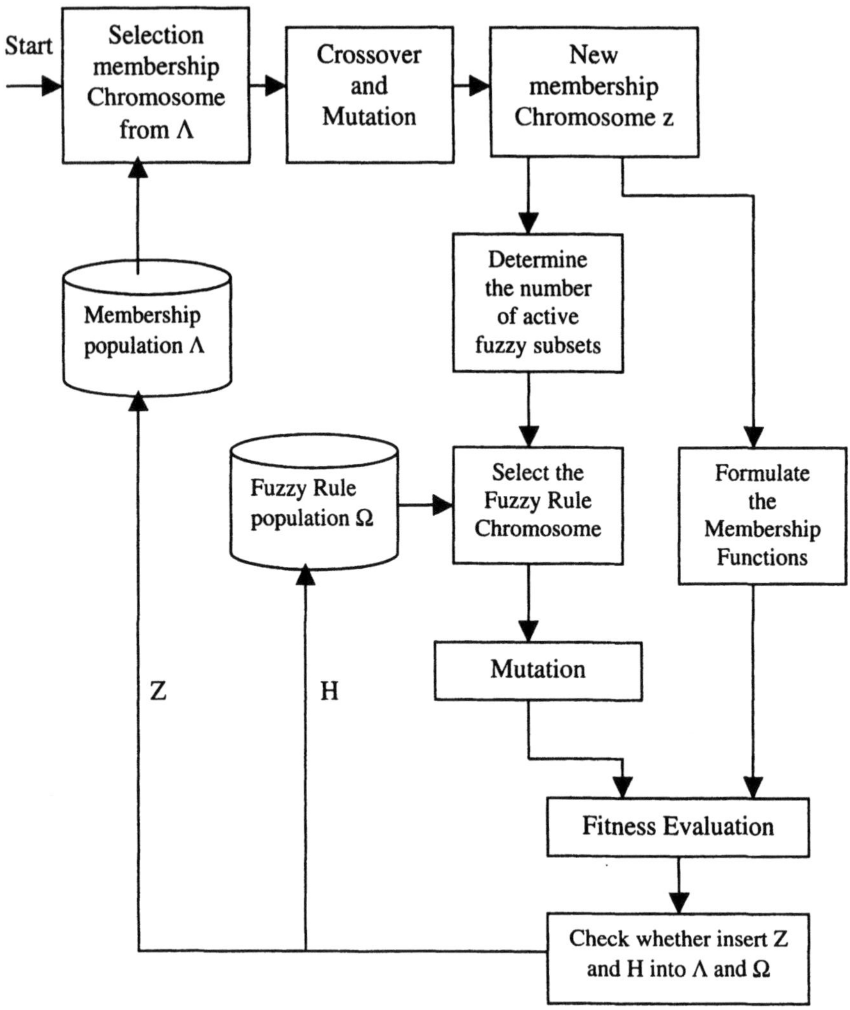

Figure 6.22 Genetic cycle for fuzzy system optimization.

There are two population pools, one for storing the membership chromosomes and the other for storing the fuzzy rule chromosomes. We can see this in Figure 6.22 as the membership population and fuzzy rule population, respectively. Considering that there are various types of gene structure, a number of different genetic operations can be used. For the crossover operation, a one-point crossover is applied separately for both the control and parameter genes of the membership chromosomes within certain operation rates. There is no crossover operation for fuzzy rule chromosomes since only one suitable rule set can be assisted.

Bit mutation is applied for the control genes of the membership chromosome. Each bit of the control gene is flipped if a probability test is satisfied (a randomly generated number is smaller than a predefined rate). As for the parameter genes, which are real number represented, random mutation is applied.

The complete genetic cycle continues until some termination criteria, for example, meeting the design specification or number of generation reaching a predefined value, are fulfilled.

6.5 Summary

We have presented in this chapter the basic concepts of genetic algorithms and simulated annealing. These optimization methodologies are motivated by nature's wisdom. Genetic algorithms emulate the process of evolution in nature, while simulated annealing is based in thermodynamic processes. We have presented classical examples of the application of these optimization techniques. We have also presented the application of genetic algorithms to the problems of optimizing neural networks and fuzzy systems. Genetic algorithms can then be viewed as a technique for efficient design of intelligent systems, because they can be used to optimize the weights or architecture of the neural network, or the number of rules in a fuzzy system. In later chapters we will make use of this fact to design intelligent systems for manufacturing in real world applications.

Chapter 7

Dynamical Systems Theory

This chapter introduces the basic concepts of dynamical systems theory, and several basic mathematical methods for controlling chaos. The main goal of this chapter is to provide an introduction to and a summary to the theory of dynamical systems with particular emphasis on fractal theory, chaos theory, and chaos control. We first define what is meant by a dynamical system, then we define an attractor, and the concept of the fractal dimension of a geometrical object. Also, we define the Lyapunov exponents as a measure of the chaotic behavior of a dynamical system. On the other hand, the fractal dimension can be used to classify geometrical objects because it measures the complexity of an object. We finish the chapter by reviewing mathematical methods for controlling chaos in dynamic systems. These methods can be used to control a real dynamic system; however, due to efficiency and accuracy requirements we will be forced to use fuzzy logic to model the uncertainty, which is present when numerical simulations are performed.

7.1 Basic Concepts of Dynamical Systems

In this section we present a brief overview of the field of Non-Linear Dynamical Systems and Fractal Theory. Recently research has shown that many simple non-linear deterministic systems can behave in an apparently unpredictable and "chaotic" manner (Grebogi, Ott, & Yorke, 1987). The existence of complicated dynamics has been discussed in the mathematical literature for many decades with important contributions by Poincaré, Birkhoft, Smale and Kolmogorov and his students, among others. Nevertheless, it is only recently that the wide-ranging impact of "chaos" has been recognized. Consequently, the field is now undergoing explosive growth, and many applications have been made across a broad spectrum of scientific disciplines-robotics, engineering, physics, chemistry, fluid mechanics and economics, to name several. We start with some basic definitions of concepts used in this chapter.

Dynamic System: This is a set of mathematical equations that allows one, in principle, to predict the future behavior of the system given the past. One example is a system of first-order ordinary differential equations in time:

$$\frac{dx}{dt} = G(x,t) \tag{7.1}$$

where x(t) is a D-dimensional vector and G is a D-dimensional vector function of x and t. Another example is a map.

Map: A map is an equation of the following form:

$$x_{t+1} = F(x_t) \tag{7.2}$$

where the "time" t is discrete and integer valued. Thus, given x_0, the maps gives x_1. Given x_1, the map gives x_2, and so on.

Dissipative system: In Hamiltonian (conservative) systems such as the ones arising in Newtonian mechanics of particles (without friction), phase space volumes are preserved by time evolution (the phase space is the space of variables that specify the state of the system). Consider, for example, a two-dimensional phase space (q, p), where q denotes a position variable and p a momentum variable. Hamilton's equations of motion take the set of initial conditions at time t $=t_0$ and evolve them in time to the set at time $t = t_1$. Although the shapes of the sets are different, their areas are the same. By a dissipative system we mean one that does not have this property. Areas should typically decrease (dissipate) in time so that the area of the final set would be less than the area of the initial set. As a consequence of this, dissipative systems typically are characterized by the presence of attractors.

Attractor: If one considers a system and its phase space, then the initial conditions may be attracted to some subset of the phase space (the attractor) as time $t \to \infty$. For example, for a damped harmonic oscillator the attractor is the point at rest. For a periodically driven oscillator in its limit cycle the limit set is a closed curve in the phase space.

Strange attractor: In the above two examples, the attractors were a point, which is a set of dimension zero, and closed curve, which is a set of dimension one. For many other attractors the attracting set can be much more irregular (some would say pathological) and, in fact, can have a dimension that is not an integer. Such sets have been called "fractal" and, when they are attractors, they are called strange attractors. The existence of a strange attractor in a physically interesting model was first demonstrated by Lorenz (see Lorenz, 1963).

Chaotic attractor: By this term we mean that if we take two typical points on the attractor that are separated from each other by a small distance $\Delta(0)$ at $t = 0$, then for increasing time t they move apart exponentially fast. That is, in some average sense:

$$\Delta(t) \sim \Delta(0) \exp(\lambda t) \tag{7.3}$$

with $\lambda > 0$ (where λ is called the Lyapunov exponent). Thus a small uncertainty in the initial state of the system rapidly leads to inability to forecast its future. It is typically the case that strange attractors are also chaotic.

One of the most prominent, chaotic, continuous-time dynamical systems is the "Lorenz attractor", named after the meteorologist E.N. Lorenz who investigated the three-dimensional, continuous-time system

$$\begin{aligned} x' &= s(-x + y) \\ y' &= rx - y - xz \qquad s, r, b > 0 \\ z' &= -bz + xy \end{aligned} \tag{7.4}$$

emerging in the study of turbulence in fluids. For r above the critical value of $r = 28.0$, trajectories of Equation (7.4) evolve in a rather unexpected way. Suppose that a trajectory starts at an initial value near the origin. For some time the trajectory regularly spirals outward from one fixed point, then the trajectory jumps to a region near another fixed point and does the same thing. As trajectories starting at different initial values all converge to and remain in the same region near the two fixed points, the region is considered an "attractor". It is a "strange attractor" because it is neither a point nor a closed curve. In general, this chaotic behavior can only occur for systems of at least three simultaneous non-linear differential equations or for systems of at least a one-dimensional non-linear map (Devaney, 1989).

Fractal geometry is a mathematical tool for dealing with complex systems that have no characteristic length scale. A well-known example is the shape of a coastline. When we see two pictures of a coastline on two different scales, we cannot tell which scale belongs to which picture: both look the same. This means that the coastline is scale invariant or, equivalently, has no characteristic length scale. Other examples in nature are rivers, cracks, mountains, and clouds. Scale-invariant systems are usually characterized by non-integer ("fractal") dimensions.

The dimension tells us how some property of an object or space changes as we view it at increased detail. There are several different types of dimension. The fractal dimension d_f describes the space filling properties of an object. Three examples of the fractal dimension are the self-similarity dimension, the capacity dimension, and the Hausdorff-Besicovitch dimension. The topological dimension d_T describes how points within an object are connected together. The embedding dimension d_e describes the space in which the object is contained.

The fractal dimensions d_f are useful and important tools to quantify self-similarity and scaling. Essentially, the dimension tells us how many new pieces are resolved as the resolution is increased. The self-similarity dimension can only be applied to geometrical self-similar objects, where the small pieces are exact copies of the whole object. However the capacity dimension can be used to analyze irregularly shaped objects that are statistically self-similar. On the other hand, the Hausdorff-Besicovitch dimension requires more complex mathematical tools. For this reason, we will limit our discussion here to the capacity dimension.

A ball is the set of points within radius r of a given point. We determine N(r) the minimum number of balls required so that each point in the object is contained within at least one ball of radius r. In order to cover all the points of the object, the balls may need to overlap. The capacity dimension is defined by the following equation:

$$d_c = \lim_{r \to 0} \frac{\log N(r)}{\log(1/r)} \ . \tag{7.5}$$

The capacity dimension defined as above is a measure of the space filling properties of an object because it gives us an idea of how much work is needed to cover the object with balls of changing size.

A useful method to determine the capacity dimension is to choose balls that are the non-overlapping boxes of a rectangular coordinate grid. N(r) is then the number of boxes with side of length r that contain at least one point of the object. Efficient algorithms have been developed to perform this "box counting" for different values of r, and thus determine the box counting dimension as the best fit of log N(r) versus log(1/r).

The fractal dimension d_f characterizes the space-filling properties of an object. The topological dimension d_T characterizes how the points that make up the object are connected together. It can have only integer values. Consider a line that is so long and wiggly that it touches every point in a plane and thus covers an area. Because it covers a plane, its space-filling fractal dimension $d_f = 2$. However, no matter how wiggly it is, it is still a line and thus has topological dimension $d_T=1$. Thus, the essence of a fractal is that its space-filling properties are larger than one anticipates from its topological dimension. Thus we can now present a formal definition of a fractal (Mandelbrot, 1987), namely, that an object is a fractal if and only if

$$d_f > d_T \, .$$

However, there is no one definition that includes all the objects or processes that have fractal properties.

Despite the identification of fractals in nearly every branch of science, too frequently the recognition of fractal structure is not accompanied with any additional insight as to its cause. Often we do not even have the foggiest idea as to the underlying dynamics leading to the fractal structure. The chaotic dynamics of non-linear systems, on the other hand, is one area where considerable progress has been made in understanding the connection with fractal geometry. Indeed, chaotic dynamics and fractal geometry have such a close relationship that one of the hallmarks of chaotic behavior has been the manifestation of fractal geometry, particularly for strange attractors in dissipative systems (Rasband, 1990). For a practical definition we take a "strange attractor", for a dynamic system, to be an attracting set with fractal dimension. For example, the famous Lorenz strange attractor has a fractal dimension of about 2.06. Also, we think that beyond only this relationship between strange attractors and the fractal dimension of the set, there is a deeper relationship between the underlying dynamics of a system and the fractal nature of its behavior. We will explore this relationship in more detail in the following sections of this chapter.

Let us consider as an example, the use of the fractal dimension as a mathematical model of the time series in the following form:

$$d = [\log(N)/\log(1/r)] \qquad (7.6)$$

where d is the fractal dimension for an object of N parts, each scaled down by a ratio r. For an estimation of this dimension we can use the following equation:

$$N(r) = \beta[\ 1/r^d\ ] \qquad (7.7)$$

where N(r) = number of boxes contained in a geometrical object and r = size of the box. We can obtain the box dimension of a geometrical object (Mandelbrot, 1987) counting the number of boxes for different sizes and performing a logarithmic regression on this data. For our particular case the geometrical object consists of the curve constructed using the set of points from the time series. We show in Figure 7.1 (a) the curve and the boxes used to cover it. In Figure 7.1 (b) the corresponding logarithmic regression is illustrated.

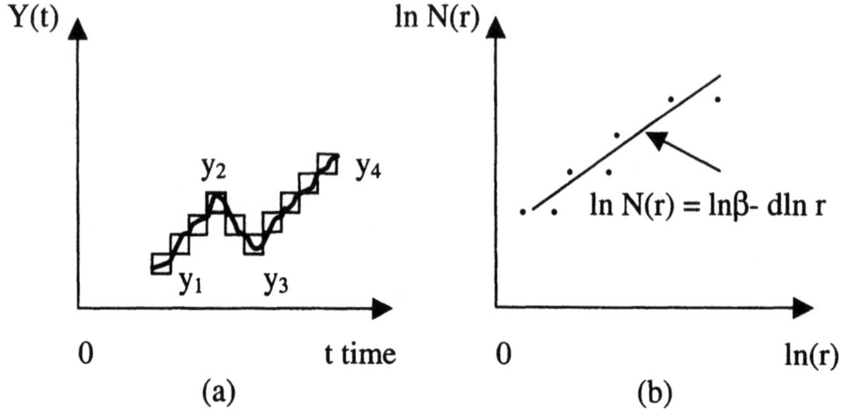

Figure 7.1 Fractal dimension of a time series: (a) curve and the boxes covering it, (b) the logarithmic regression to find d

7.2 Controlling Chaos

More than two decades of intensive studies on non-linear dynamics have posed the question on the practical applications of chaos (Kapitaniak, 1996). One of the possible answers is to control chaotic behavior in such a way as to make it predictable. Indeed, nowadays the idea of controlling chaos is an appealing one.

Chaos occurs widely in engineering and natural systems; historically it has usually been regarded as a nuisance and is designed out if possible. It has been noted only as irregular or unpredictable behavior, often attributed to random external influences. More recently, there have been examples of the potential usefulness of chaotic behavior (Kapitaniak, 1996).

We can divide chaos controlling approaches into two broad categories: firstly those in which the actual trajectory in the phase space of the system is monitored and some "feedback" process is employed to maintain the trajectory in the desired mode, and secondly "non-feedback" methods in which some other property or knowledge of the system is used to modify or exploit chaotic behavior. Feedback methods do not change the controlled systems and stabilize unstable periodic orbits or strange chaotic attractors, while non-feedback methods slightly change the controlled system, mainly by a small permanent shift of control parameter, changing the system behavior from chaotic attractor to periodic orbit which is close to the initial attractor.

We describe in this section several methods by which chaotic behavior in a dynamical system may be modified, displaced in parameter space or removed.

The Ott-Grebogi-Yorke (OGY) method (Ott et al.,1990) is extremely general, relying only on the universal property of chaotic attractors, namely that they have embedded within them infinitely many unstable periodic orbits. On the other hand, the method requires following the trajectory and employing a feedback control system, which must be highly flexible and responsive; such a system in some experimental configurations may be large and expensive. It has the additional disadvantage that small amounts of noise may cause occasional large departures from the desired operating trajectory.

The non-feedback approach is inevitable much less flexible, and requires more prior knowledge of equations of motion. On the other hand, to apply such a method, we do not have to follow the trajectory. The control procedures can be applied at any time and we can switch from one periodic orbit to another without returning to the chaotic behavior, although after each switch, transient chaos may be observed. The lifetime of this transient chaos strongly depends on initial conditions. Moreover, in a non-feedback method we do not have to wait until the trajectory is close to an appropriate unstable orbit; in some cases this time can be quit long. The dynamic approach can be very useful in mechanical systems, where feedback controllers are often very large. In contrast, a dynamical absorber having a mass of the order of 1% of that of the control system is able, as we will see later, to convert chaotic behavior to periodic one over a substantial region of parameter space. Indeed, the simplicity by which chaotic behavior may be changed in this way may actually motivate the search for, and exploitation of, chaotic behavior in practical systems.

The essential property of a chaotic trajectory is that it is not asymptotically stable. Closely correlated initial conditions have trajectories, which quickly become uncorrelated. Despite this obvious disadvantage, it has been established that control leading to the synchronization of two chaotic systems is possible.

The methods described in this section are illustrated by the example of Chua's circuit (Chua, 1993) shown in Figure 7.2. Chua's circuit contains three linear energy storage elements (an inductor and two capacitors), a linear resistor, and a single non-linear resistor N_R, namely Chua's diode with a three segment piecewise linear v-i characteristic defined by

$$f(v_{c1})=m_0v_{c1} + \tfrac{1}{2}(m_1 - m_0)(|v_{c1} + 1|-| v_{c1} - 1|) \qquad (7.8)$$

where the slopes in the inner and outer regions are m_0 and m_1 respectively (this is shown in Figure 7.3).

In this case the state equations for the dynamics of Chua's circuit are as follows:

$$C_1\frac{dv_{c1}}{dt}= G(v_{c2} - v_{c1}) - f(v_{c1})$$

$$C_2\frac{dv_{c2}}{dt} = G(v_{c1} - v_{c2}) - i_L \qquad (7.9)$$

$$L\frac{di_L}{dt} = v_{c2}$$

where G=1/R.

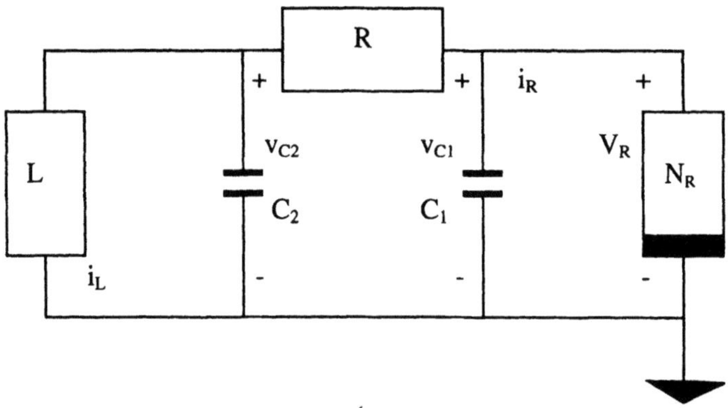

Figure 7.2 Chua's circuit.

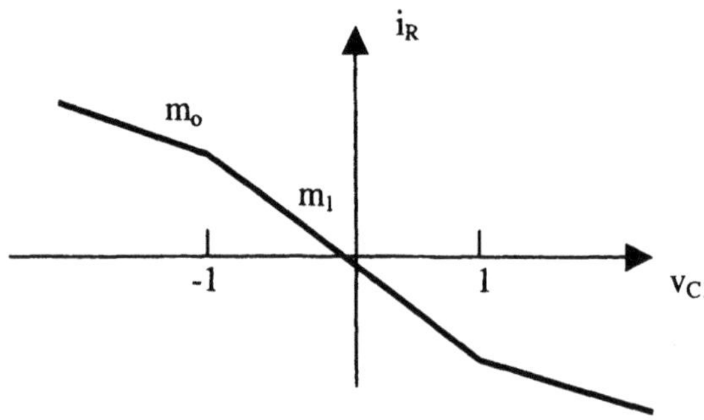

Figure 7.3 i_R-V_{c1} characteristic of the non-linear resistor.

It is well-known that for R = 1.64 kΩ, C_1 = 10 nF, C_2 = 99.34 nF, m_1 = -0.76 mS, m_0 = 0.41 mS, and L = 18.46 mH, Chua's circuit operate on the chaotic double-scroll Chua's attractor. We show in the following figures the simulation of Chua's circuit for initial conditions (-3, -3, -10). Figure 7.4 shows the plot of variable Vc_1

in time. In this figure, we can appreciate the erratic behavior of this variable. Figure 7.5 shows the plot of variable Vc_2 across time, which is similar to the behavior of Vc_1. We also show in Figure 7.6 the plot of variable i_L in time. In this figure, we can appreciate the erratic fluctuations of the current intensity across time. In Figure 7.7 we show a bi-dimensional view of the double-scroll Chua's attractor. Finally, in Figure 7.8 we can appreciate a three-dimensional view of Chua's attractor.

The chaotic dynamics of Chua's circuit have been widely investigated (e.g. Madan, 1993). One of the main advantages of this system is the very good accuracy between numerical simulations of the model and experiments on real electronic devices. Experiments with this circuit are very easy to perform, even for non-specialists.

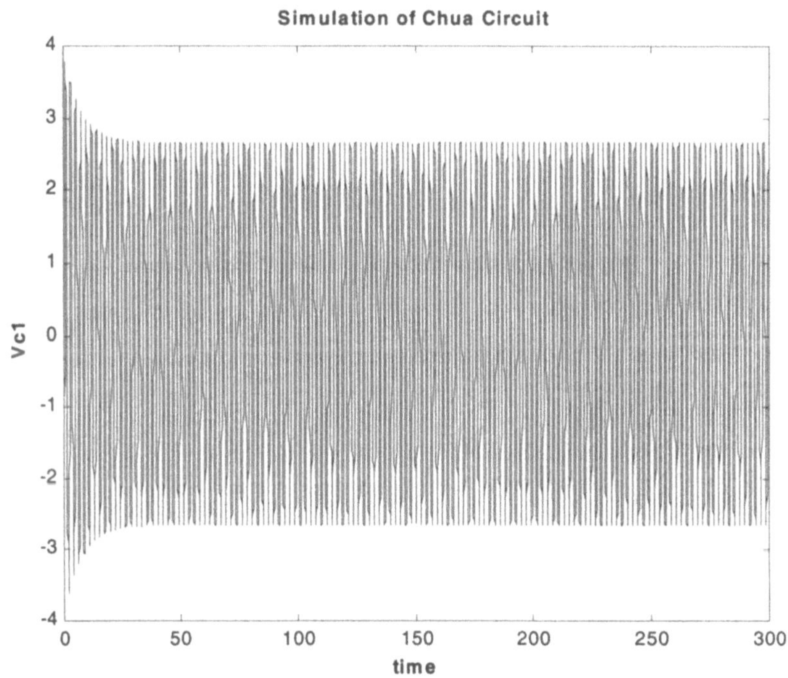

Figure 7.4 Plot of variable Vc_1 of Chua's circuit.

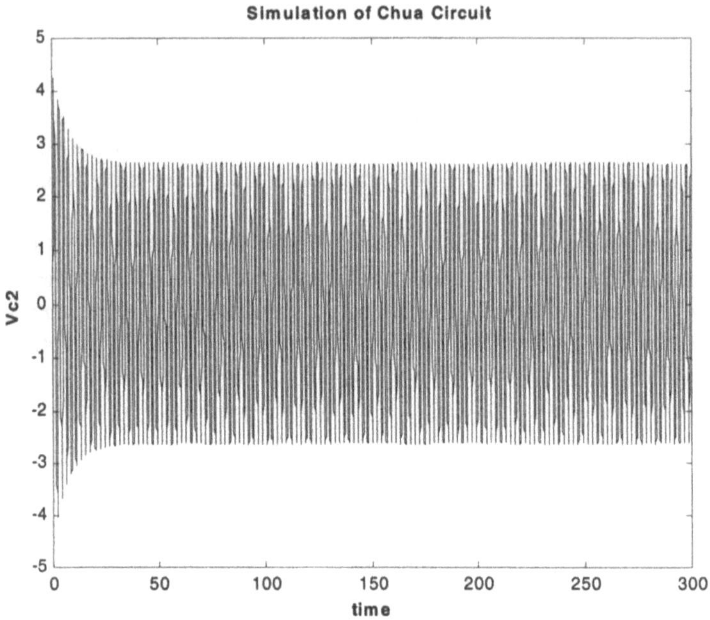

Figure 7.5 Plot of variable Vc2 for Chua's circuit.

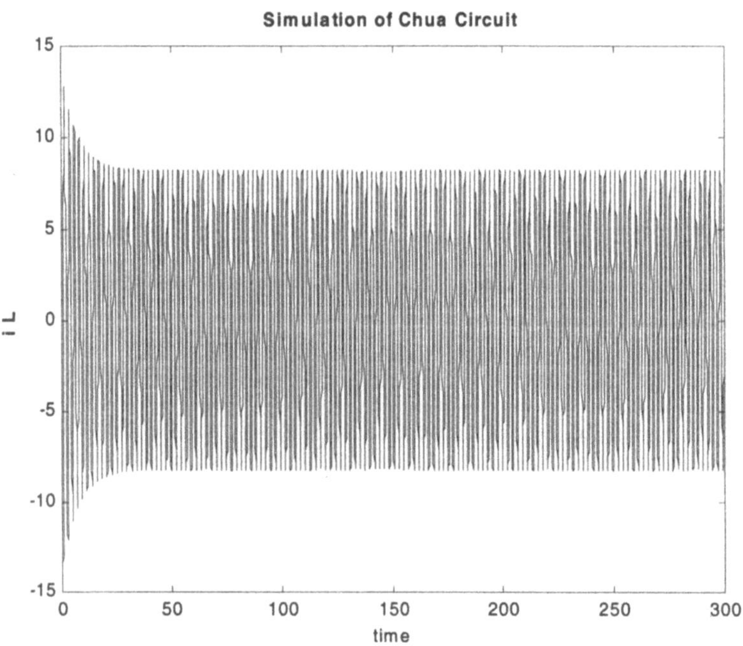

Figure 7.6 Plot of variable i_L for Chua's circuit.

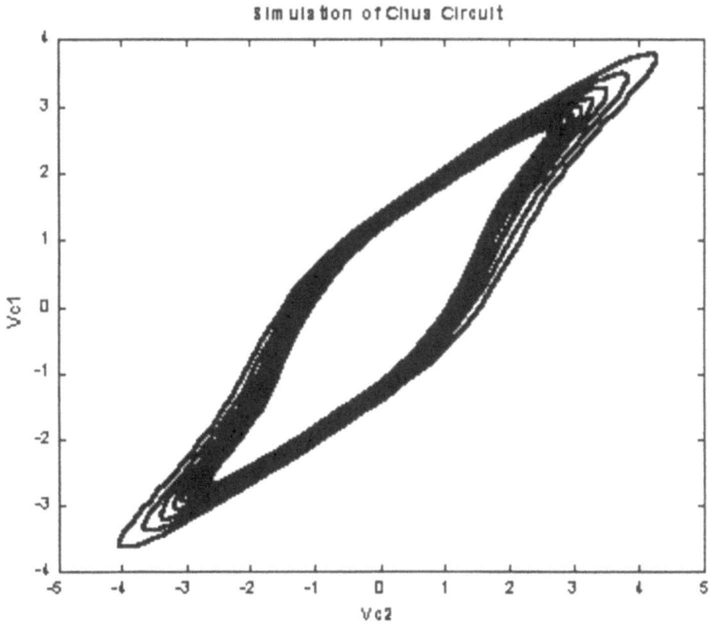

Figure 7.7 Bi-dimensional view of Chua's attractor.

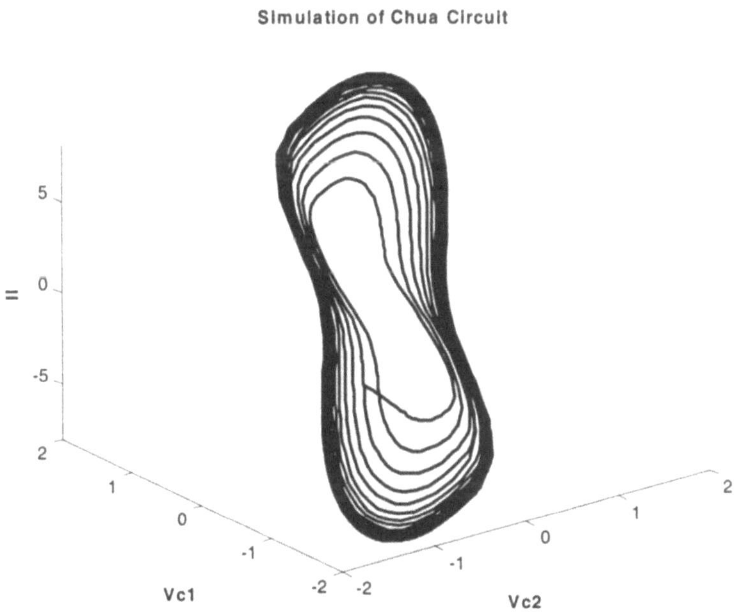

Figure 7.8 Three-dimensional view of Chua's attractor.

7.2.1 Controlling Chaos through Feedback

7.2.1.1 Ott-Grebogi-Yorke Method

Ott, Grebogi and Yorke (Ott et al., 1990) have proposed an developed a method by which chaos can always be suppressed by shadowing one of the infinitely many unstable periodic orbits embedded in the chaotic attractor.

The basic assumptions of this method are as follows.

(a) The dynamics of the system can be described by an n-dimensional map of the form:

$$X_{n+1} = f(X_n, p) \tag{7.10}$$

(b) p is some accessible system parameter, which can be changed in some small neighborhood of its nominal value p*.
(c) For this value p* there is a periodic orbit within the attractor around which we would like to stabilize the system.
(d) The position of this orbit changes smoothly with changes in p, and there are small changes in the local system behavior for small variations of p.

Let X_F be a chosen fixed point of the map f of the system existing for the parameter value p*. In the close vicinity of this fixed point with good accuracy we can assume that the dynamics are linear and can be expressed approximately by

$$X_{n+1} - X_F = M(X_n - X_F) \tag{7.11}$$

The elements of the matrix M can be calculated using the measured chaotic time series and analyzing its behavior in the neighborhood of the fixed point. The OGY algorithm is schematically explained in Figure 7.8 and its main properties are as follows.

(a) No model of dynamics is required. One can use either full information from the process or a delay coordinate embedding technique using single variable experimental time series.
(b) Any accessible variable (controllable) system parameter can be used as the control parameter.
(c) In the absence of noise and error, the amplitude of applied control signal must be large enough (exceed a threshold) to achieve control.
(d) Inevitable noise can destabilize the controlled orbit, resulting in occasional chaotic bursts.
(e) Before settling into the desired periodic mode, the trajectory exhibits chaotic transients, the length of which depends on the actual starting point.

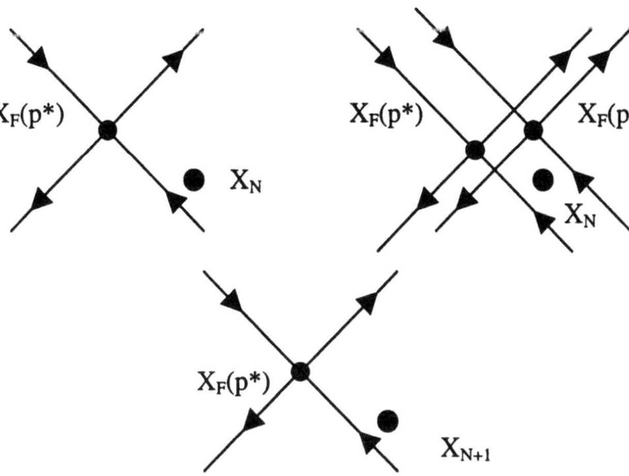

Figure 7.9 General idea of the Ott-Grebogi-Yorke method.

In Ogorzalek (1993) the OGY method was applied to control chaos in Chua's circuit. Using a specific software package, unstable periodic orbits embedded in the attractor, which could serve as goals of control were found. The controlling method was implemented in the way shown in Figure 7.9. The computer was used for data acquisition, identification of the chaotic system in terms of unstable periodic orbits and calculation of the control signal. When applying the OGY method to control chaos in a real electronic circuit the main problem encountered was the noise introduced due to inevitable noise of the circuit elements. The method was found to be very sensitive to the noise level – very small signals sometimes are hidden within the noise, and control is impossible.

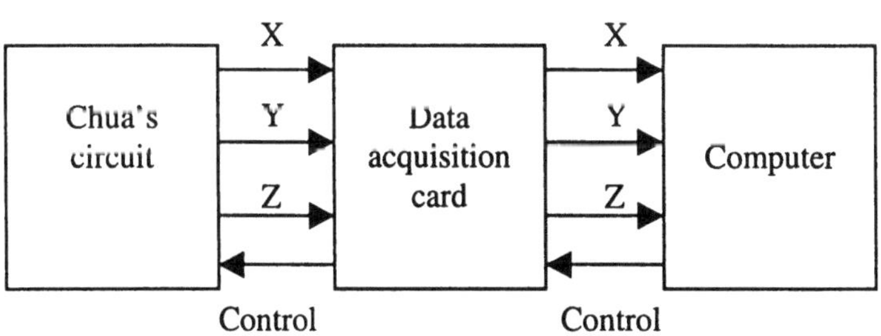

Figure 7.10 Practical implementation of OGY method.

Generally, the experimental application of the OGY method requires a permanent computer analysis of the state of the system. The changes of the parameters, however, are discrete in time and this leads to some serious limitations. The method can stabilize only those periodic orbits which maximal Lyapunov exponent is small compared to the reciprocal of the time interval between parameter changes. Since the corrections of the parameter are rare and small, the fluctuation noise leads to occasional bursts of the system into regions far from the desired periodic orbit, especially in the presence of noise.

7.2.1.2 Pyragas's Control Methods

A different approach to feedback control which allows the above mentioned problems to be avoided was proposed by Pyragas (1992). This method is based on the construction of a special form of a time continuous perturbation, which does not change the form of the desired unstable periodic orbit, but under certain constraints can stabilize it. Two feedback controlling loops, shown in Figure 7.10, have been proposed.

A combination of feedback and periodic external force is used in the first method (Figure 7.10(a)). The second method (Figure 7.10(b)) does not require any external source of energy and it is based on self-controlling delayed feedback. If the period of external force or a time delay is equal to the period of one of the unstable periodic orbits embedded in the chaotic attractor it is possible to find a constant K which allows stabilization of the unstable periodic orbit. This approach, being noise resistant, can easily be used in experimental systems. The first of Pyragas's methods can be considered as the special case of the direct application of classical controlling methods to the problem of controlling chaos.

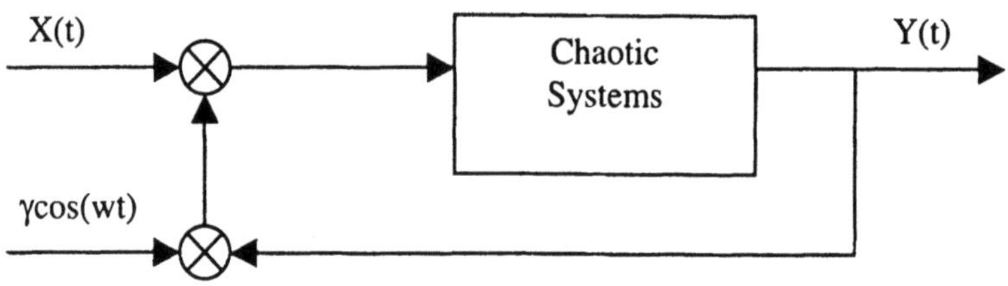

(a)

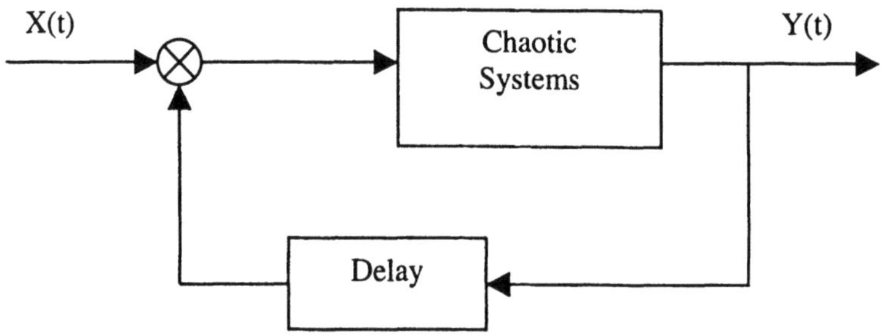

(b)

Figure 7.11 Feedback controlling loops; (a) control by periodic external perturbation, (b) control by time delay.

The dynamical system

$$X' = f(X) \tag{7.12}$$

Where $x \in R^n$, is controllable if there exists a control function u(t), such that

$$X' = f(X) + u(t) \tag{7.13}$$

allows to move trajectory X(t) from point X_0 at time t_0 to the desired point X in finite time T.

The controllability concept can be applied to the chaos controlling problems. For example, for Chua's circuit the equations for the controlled circuit are

$$\begin{aligned} X' &= a(Y-X-f(X)) \\ Y' &= X-Y+Z-K(Y-Y^*) \\ Z' &= -bY \end{aligned} \tag{7.14}$$

This approach is illustrated in Figure 7.11. The main advantages of this method are as follows.

(a) Any solution of the original system can be a goal of the control (fixed point, unstable periodic orbit, etc.)
(b) The controller has a very simple structure.
(c) Access to system parameters is not required.
(d) It is not affected by small parameter variations.

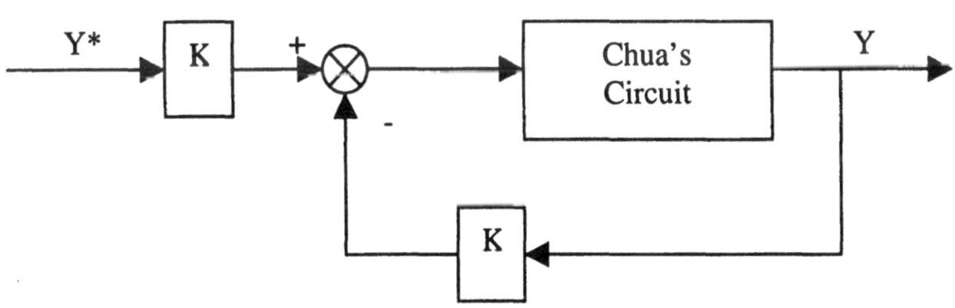

Figure 7.12 Closed loop feedback control configuration.

7.2.1.3 Controlling Chaos by Chaos

In this section, we show that the chaotic behavior of one system can be controlled by coupling it with another one which can also be chaotic (Kapitaniak, 1996). Thus we consider two chaotic systems, which we call A and B respectively,

$$X' = f(X)$$
$$Y' = g(X) \qquad (7.15)$$

where $x,y \in R^n$, and we use the controlling strategy which is schematically illustrated in Figure 7.12; the two systems are coupled through the operators λ,μ, which have a very simple linear form. We assume that some or all state variables of both systems A and B can be measured, so that we can measure signal X(t) from system A and signal Y(t) from B, and that the systems are coupled in such a way that the differences D_1 and D_2 between the signals X(t) and Y(t) are

$$F_1(t) = \lambda[X(t)-Y(t)] = \lambda D_1(t)$$
$$F_2(t) = \mu[Y(t)-X(t)] = \mu D_2(t) \qquad (7.16)$$

used as control signals introduced respectively into each of the chaotic systems A and B as negative feedback. We take $\lambda,\ \mu > 0$ to be experimentally adjustable weights of the perturbation.

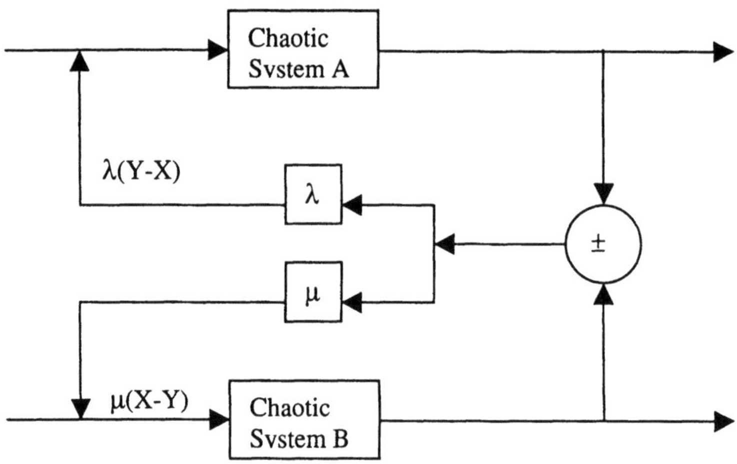

Figure 7.13 Controlling chaos by chaos scheme.

Using the coupling schematically shown in Figure 7.12, it has been shown that one chaotic system coupled with the other one can significantly change the behavior of one of them (unidirectional coupling, i.e. λ or $\mu = 0$) or of both systems (mutual coupling, i.e. $\lambda,\ \mu \neq 0$). This property allows us to describe the above procedure as the "controlling chaos by chaos" method. In Kocarev and Kapitaniak (1995), rigorous conditions are given, under which chaotic attractors of systems A and B are equivalent, or the evolution of one of them is forced to take place on the attractor of the other one. Kapitaniak (1996) shows an example of coupling two Lorenz chaotic attractors, which results in chaos control and increase of the predictability.

7.2.2 Controlling Chaos without Feedback

7.2.2.1 Control through Operating Conditions

Virtually all engineering and most natural systems are subjected during operation to external forcing. This forcing will contain (and hopefully be dominated by) planned and intentional components; it will also almost invariably contain unintentional "noise". Smart design and control of this forcing is often able to annihilate, or shift to a harmless region of parameter space, an unwanted chaotic behavior.

In this case, the method consists in finding the chaotic region in parameter space by analytical and numerical methods (Kapitaniak, 1996). Then based on this region change the parameters to control the dynamical system. The procedure described in this section is based on the direct change of one of the system parameters to shift system behavior from chaotic to periodic, close to the chaotic attractor. It cannot be called a control method in the sense of the methods described before, but it illustrates that having a system designed as chaotic, we obtain easy access to different types of periodic behavior.

7.2.2.2 Control by System Design

In this section, we explore the idea of modifying or removing chaotic behavior by appropriate system design. It is clear that, to a certain extent, chaos may be "designed out" of a system by appropriate modification of parameters, perhaps corresponding to modification of mass or inertia of moving parts. Equally clearly, there exist strict limits beyond which such modifications cannot go without seriously affecting the efficiency of the system itself.

In this section, we describe a method for controlling chaos in which the chaos effect is achieved by coupling the chaotic main system to a simpler autonomous system (controller), usually linear, as shown in Figure 7.13. This method (Kapitaniak, 1996) is developed for chaotic systems in which for some reason it is difficult, if not impossible, to change any parameter of the main system. In particular consider the coupling of the chaotic system

$$X' = f(X,\mu) \tag{7.17}$$

where $x \in R^n$, $n \geq 3$ and $\mu \in R$ is a system parameter, to another (simpler) asymptotically stable system (controller) described by

$$Y' = g(Y,e) \tag{7.18}$$

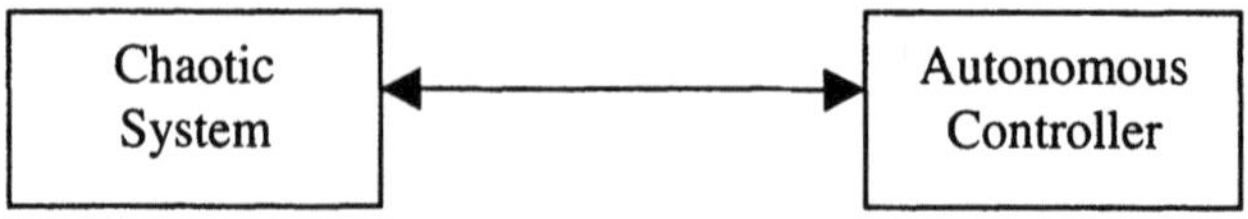

Figure 7.14 Coupling scheme.

where $Y \in R^n$, e is a vector denoting the controller's parameters, where at least one of the parameters e_i can be easily changed. For practical reasons, the dimension m of the controller system (7.18) should be chosen as low as possible. Since the method was mainly designed for controlling chaos in mechanical systems, we choose m = 2, i.e., a one degree of freedom controller (the simplest mechanical system). The equations for the extended system are

$$\begin{aligned} X' &= f(X,\mu) + AY \\ Y' &= g(Y,e) + BX \end{aligned} \tag{7.19}$$

where A and B are the coupling matrices. Since the Y subsystem is asymptotically stable, the role of the controller is to change the behavior of the system from chaotic to some desired periodic, possibly constant, operating regime.

The idea of this method is similar to that of the so-called dynamical vibration absorber. A dynamical vibration absorber is a one degree of freedom system, usually mass on a spring, which is connected to the main system as shown in Figure 7.14.

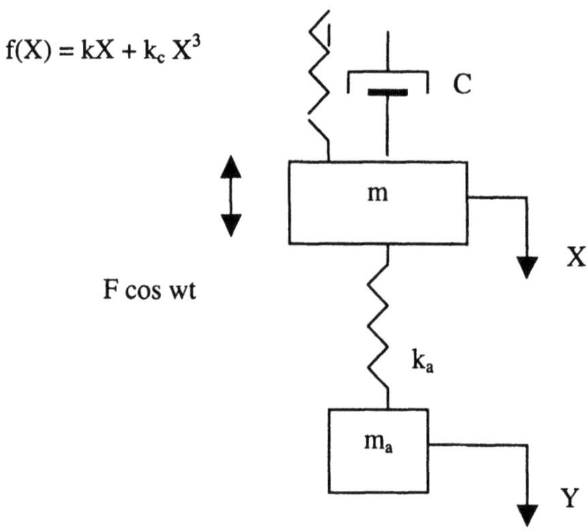

Figure 7.15 Dynamical damper as chaos controller.

Although such a dynamical absorber can change the overall dynamics substantially, it need usually only be physically small in comparison with the main system, and does not require an increase of excitation force. It can be easily added to the existing system without major changes of design or construction. This contrasts with devices based on feedback control, which can be large and costly.

To explain the role of dynamical absorbers in controlling chaotic behavior let us consider the Duffing oscillator, coupled with an additional linear system:

$$X'' + aX' + bX + cX^3 + d(X-Y) = B_0 + B_1 \cos wt \qquad (7.20a)$$

$$Y'' + e(Y-X) = 0 \qquad (7.20b)$$

where a, b, c, d, e, B_0, B_1, and w are constants. Here d and e are the characteristic parameters for the absorber, and we take e as the control parameter.

It is well known that the Duffing's oscillator shows chaotic behavior for certain parameter regions. As has been mentioned in the previous section, in many cases the route to chaos proceeds via s sequence of period doubling bifurcations, and in such cases this method provides an easy way of switching between chaotic and periodic behavior. We show in Figure 7.15 a two-dimensional view of the chaotic behavior in Duffing's oscillator. In Figure 7.16 we show a plot of variable X across time [0, 350]. In Figure 7.17 we show a plot of variable X' across time. Finally, we show in Figure 7.18 a three-dimensional view of the strange attractor.

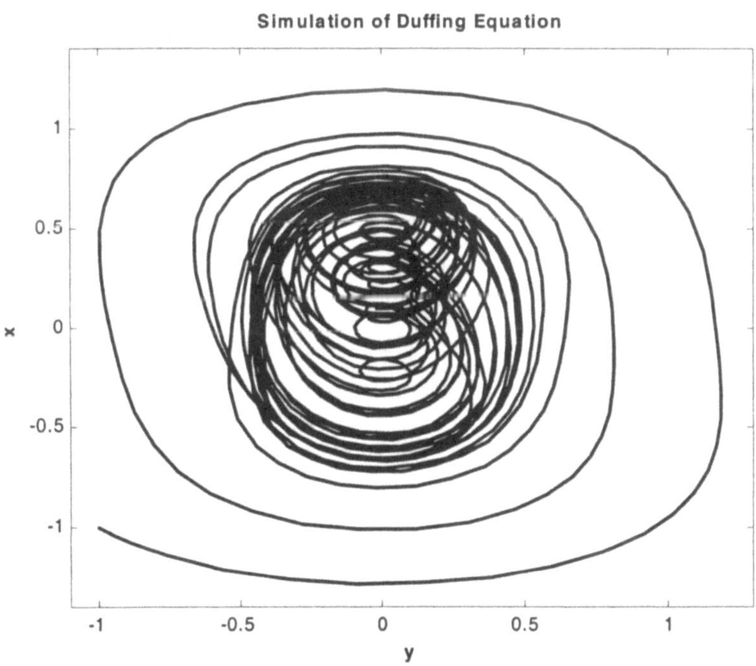

Figure 7.16 Chaotic behavior in Duffing's oscillator.

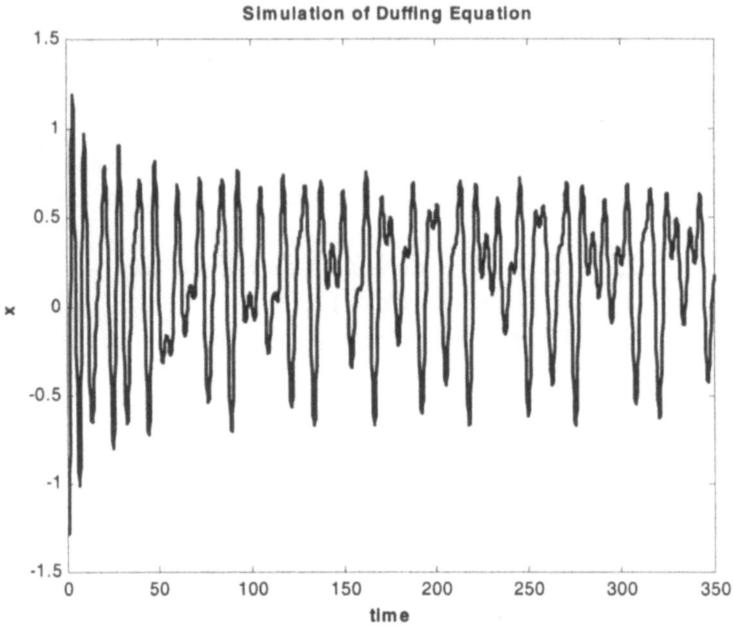

Figure 7.17 Plot of variable X across time for Duffing's oscillator.

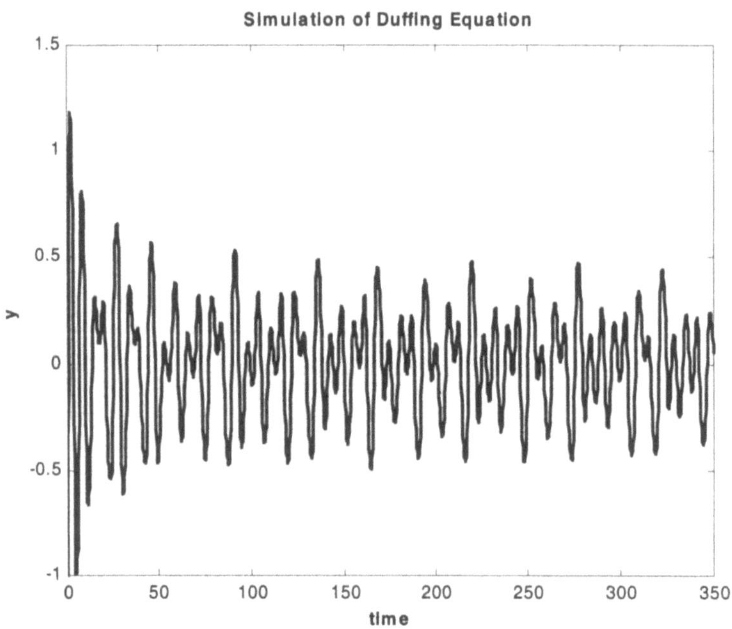

Figure 7.18 Plot of variable X' across time for Duffing's oscillator.

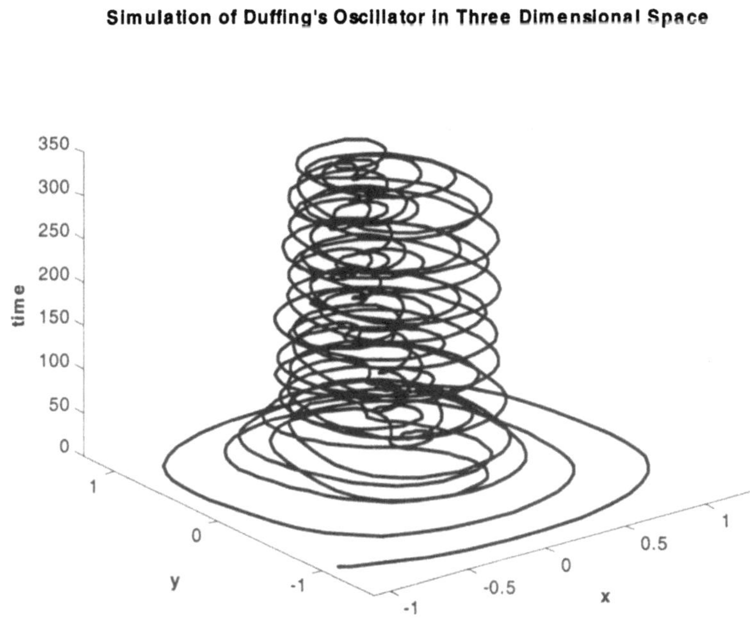

Figure 7.19 Simulation of Duffing's oscillator (Equation 7.20) for e = 0.01.

Let us consider the parameters of Equation (7.20) to be fixed at the values a=0.077, b=0, c=1.0, B_0=0.045, B_1=0.16, w=1.0, then we can find (Kapitaniak, 1996) that we have chaos for e $\in$ [0, 0.10], and we can control this chaos by increasing e above 0.10. As this method is designed mainly for experimental applications, we shall now briefly suggest some guidelines for applying it.

(1) The coupled system has to be as simple as possible.
(2) The coupling e should be chosen as small as possible.
(3) If it is possible one should couple the controller in such a way that the locations of the fixed points of the original system are not changed.

7.2.2.3 Taming Chaos

In paper by Steeb et al. (1986) it was first demonstrated that chaos in a dynamical system can be reduced (the largest Lyapunov exponent is decreased) or replaced by regular behavior by applying a weak external periodic signal. Periodic perturbation can be introduced to the system as external force or as a perturbation of one of the internal system parameters. Given an external perturbation, it is possible to show that a chaotic system is capable of finding an appropriate orbit.

In Kapitaniak (1988) it was shown that the chaotic system can be set into regular motion by addition of suitable random noise. Other authors have described similar approaches. Later these approaches have been called "taming chaos". Possible outputs of the taming chaos procedure are shown in Figure 7.19.

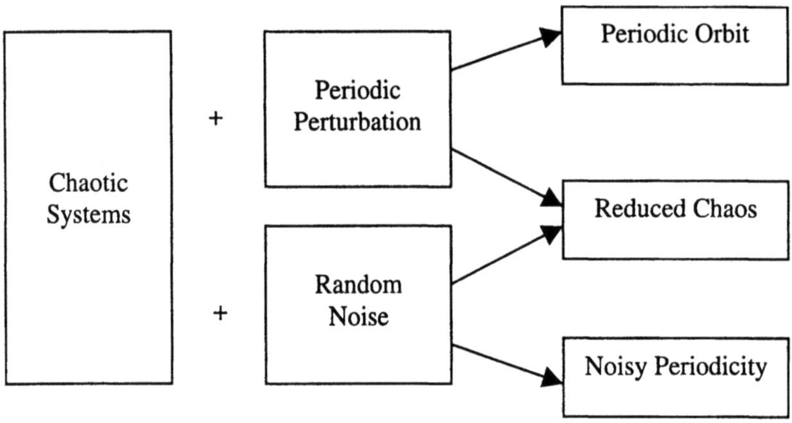

Figure 7.20 General idea of taming chaos.

7.2.3 Method Selection

Although the methods described in the previous sections have been developed mainly by physicists and mathematicians, generally most of them can be applied to control engineering systems.

In particular, the non-feedback methods can practically always be used. Their applications are straightforward and do not require special complicated controllers to be used. The main disadvantage of these methods is that the goal of controlling has to be determined by trial and error method.

The motivations for using feedback systems to control chaos are the following: feedback controllers are easy to implement, especially in electrical systems, they can perform the job automatically, and stabilize the overall control system efficiently. On the other hand, conventional feedback controllers are designed for non-chaotic systems. A chaotic system sensitivity to initial conditions may lead to the impression that in chaotic systems their sensitivity to small errors makes them very difficult. Such an impression may lead to the argument that once the control is initiated there is no need for further monitoring of the system's dynamics, nor feeding back this information in order to sustain the control. Indeed, it turns out that conventional feedback control of chaotic systems is generally difficult, but not impossible. Recently, Chen and Dong (1995) used a neural network approach for identification and control of chaotic systems. In many cases, a specially implemented feedback method can guarantee stabilization of the dynamic system. To summarize, the selection of the controlling method has to be based on the following:

(1) the goal of controlling (e.g. if the suppression of chaos is the main goal, then non-feedback methods can be applied in an easier way);
(2) the level of noise in the system (e.g. if the level of noise is large, then Pyragas's methods can be more effective than the OGY approach);
(3) the particular characteristics of the system. (Generally, in electrical systems one can try to use both feedback and non-feedback methods. In mechanical systems where the suppression of chaos is the main goal of controlling, non-feedback methods are recommended).

7.3 Summary

In this chapter, we have presented the main ideas underlying Dynamical Systems, Fractal theory, and Chaos theory, and we have only started to point out the many possible applications of these powerful mathematical theories. We have discussed in some detail the concepts of strange attractors, chaotic behavior and fractal dimension. The concept of the fractal dimension will be the basis of the method for time series analysis that will be described later in this book, which will be used to achieve plant monitoring and diagnostics and also in automated quality control. Also, we have introduced several methods for controlling chaos that use some of the basic concepts of dynamical systems. These methods for controlling chaos can be used for real time control of dynamical systems, or can be used in combination with other computational techniques when the complexity of the problem requires more efficiency and accuracy. We will see in a later chapter that many real world problems require hybrid control systems that combine several techniques to achieve the desired level of performance in manufacturing systems.

Chapter 8

Plant Monitoring and Diagnostics

We describe in this chapter a new hybrid fuzzy-fractal approach for plant monitoring and diagnostics. We use the concept of the fractal dimension to measure the complexity of a time series of observed data from the plant. We also use fuzzy logic to represent expert knowledge on monitoring the process in the plant. In the hybrid fuzzy-fractal approach a set of fuzzy if-then rules are used to classify different conditions of the plant. The fractal dimension is used as input linguistic variable in the fuzzy system to improve the accuracy in the classification. An implementation of the proposed approach is shown to describe in more detail the method.

Diagnostic systems are used to monitor the behavior of a process and identify certain pre-defined patterns that are associated with well-known problems (Du, 1998). These problems, once identified, imply suggestions for specific solutions. Some diagnostic systems are in the form of a rule-based expert system: a set of rules is used to describe certain patterns (Chiang, Russell & Braatz, 2000). Observed data are collected and used to evaluate these rules. If the rules are logically satisfied, the pattern is identified, and the problem associated with that pattern is suggested. In general, the diagnostic systems are used for consultation rather than replacement of human expert.

Most current plant monitoring systems only check a few variables against individual upper and lower limits, and start an audible alarm should each variable move out of its predefined range (Chen & Pham, 2001). Other more complicated systems normally involve more sensors that provide more data but still follow the same pattern of independently checking individual sets of data against some upper and lower limits. The warning alarm from these systems only carries a meaning that there is something wrong with the process in the plant.

In this chapter a new fuzzy-fractal approach for plant monitoring is proposed. The concept of the fractal dimension is used to measure the complexity of the time series of relevant variables for the process (Castillo & Melin, 1994), (Mandelbrot, 1997). The fractal dimension is a mathematical concept that basically measures how much work is needed to cover a geometrical object in space (Peitgen, Saupe, Jurgens & Yunker, 1992). For the case of monitoring, the fractal dimension is used to extract information from the time series of the relevant variables (Tricot, 1994). A set of fuzzy rules is used to represent the knowledge for monitoring the process (Du, Elbestawi & Wu, 1993). In the fuzzy rules, the fractal dimension is used as a linguistic variable (Zadeh, 1973) to help in recognizing specific patterns in the measured data. The fuzzy-fractal approach has been applied before in problems of financial time series prediction (Castillo & Melin, 1996, 1998, 2001) and for other types of problems (Melin & Castillo, 1998), but now it is proposed to the monitoring of plants.

This chapter is structured as follows. First, the problem of plant monitoring and diagnosis is described in more detail. Second, some basic concepts of fractal theory are given. Third, the problem of fuzzy modeling for monitoring and diagnosis is described in more detail. Then, the fuzzy-fractal approach for plant monitoring is proposed and explained. Finally, some experimental results are given and future work is proposed.

8.1 Monitoring and Diagnosis

Monitoring means checking or regulating the performance of a machine, a process, or a system (Chiang, Russell & Braatz, 2000). Diagnosis, on the other hand, means deciding the nature and the cause of a diseased condition of a machine, a process, or a system by examining the symptoms. In other words, monitoring is detecting suspect symptoms, whereas diagnosis is determining the cause of the symptoms. There are several words and phrases that have similar or slightly different meanings such as fault detection, fault prediction, in-process evaluation, on-line inspection, identification, and estimation (Patton, Frank & Clark, 2000). The problems of engineering monitoring and diagnosis vary. The following list gives a few examples:

1) monitoring the deterioration of a high-pressure vessel based on acoustic emission signals;
2) monitoring the tool condition in a transfer machining station based on motor current signals;
3) diagnosing the condition of a turbine machinery set based on vibration signals.

The importance of monitoring and diagnosis of plant processes now is widely recognized because it results in increased productivity, improved product quality and decreased production cost (Du, 1998). As a result, in the past decade, a large number of research and development projects have been carried and many monitoring and diagnosis methods have been developed (Du, Elbestawi & Wu, 1993). The commonly used monitoring and diagnosis methods include modeling-based methods (Russell, Chiang, & Braatz, 2000), pattern recognition methods (Scharf, 1991), fuzzy systems methods (Mendel, 2001), knowledge-based systems methods (Wang, 1997), and artificial neural networks (Jain, et.al., 2000). It is interesting to note that even though these methods are rather different, they share a very similar structure as shown in Figure 8.1.

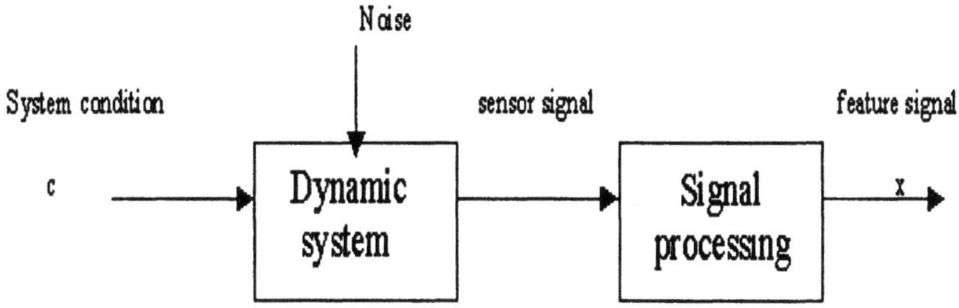

Figure 8.1. Plant Monitoring and Diagnosis.

In Figure 8.1 the engineering system is considered to be dynamic, i.e. changing its condition with time. For this reason, there is a block in Figure 8.1 called "dynamic system". Also, in Figure 8.1 there is block indicating that "signal processing" has to be performed on the sensor signals to extract the feature signals, which in turn are used to be able to monitor the dynamic process. Of course, Figure 8.1 only shows the general structure for engineering monitoring and diagnosis.

The "condition" of a machine, a process, or an engineering system (which will be referred to as system condition and denoted by $c \in \{c_1, c_2, \ldots, c_m\}$) can be considered as the "input", the system working conditions and noises (including system noise and sampling noise) can be considered as the "noise", and the sensor signals are the "outputs" from the system. Typically, the sensor signals are processed by a computer, after which the signals are transformed into a set of features called feature signals, denoted as $\mathbf{x} = \{x_1, x_2, \ldots, x_n\}$. In general, the systems conditions are predefined, such as normal, critical, etc. On the other hand, the features may be the mean of a temperature signal, the variance of a displacement signal, etc. Sensing and signal processing are very important to the success of plant monitoring and diagnosis (Chen & Pham, 2001).

More formally, the goal of monitoring is to use the feature signals, x, to determine whether the plant is in an acceptable condition(s) (a subset of $\{c_1, c_2, \ldots, c_m\}$). On the other hand, the objective of diagnosis is to use the feature signals, x, to determine the system condition, $c \in \{c_1, c_2, \ldots, c_m\}$. No matter how monitoring and diagnosis methods may differ, monitoring and diagnosis always consist of two phases: training and decision making. Training is to establish a relationship between the feature signals and the systems conditions. Without losing generality, this relationship can be represented as

$$\mathbf{x} = F(c). \tag{8.1}$$

It should be pointed out that F(c) represents a fuzzy system, a neural network or another method that could be used to obtain this relationship (Melin & Castillo,

2002). In fact, it is the form of the relationship that determines the methods of monitoring and diagnosis, as well as the performance of the methods. The relationship F(c) is established based on training samples, denoted by $\mathbf{x}_1$, $\mathbf{x}_2$,..., $\mathbf{x}_k$,..., $\mathbf{x}_N$, where the system condition for each training sample is known [and denoted as c($\mathbf{x}_k$)].

After the relationship is established, when a new sample is given (from an unknown system condition), its corresponding condition is estimated based on the inverse relationship

$$c = F^{-1}(\mathbf{x}). \tag{8.2}$$

This is called decision-making, or classifying. Whereas it is not likely that the training samples will cover all possible cases, decision making often involves reasoning or inferencing. Of course, if the inverse relationship doesn't exist then a "pseudo-inverse" approach can be used instead.

In particular, when a fuzzy system is used, the relationship is given by a set of rules as shown in Figure 8.2. The input to the fuzzy system is the feature signal and the output of the fuzzy system is the estimated plant condition(s) [i.e., $\mathbf{z} = (z_1, z_2,..., z_m)$ is an estimate of $\mathbf{c} = (c_1, c_2,..., c_m)$]. In other words, the fuzzy system models the inverse relationship between the system conditions and the feature signals.

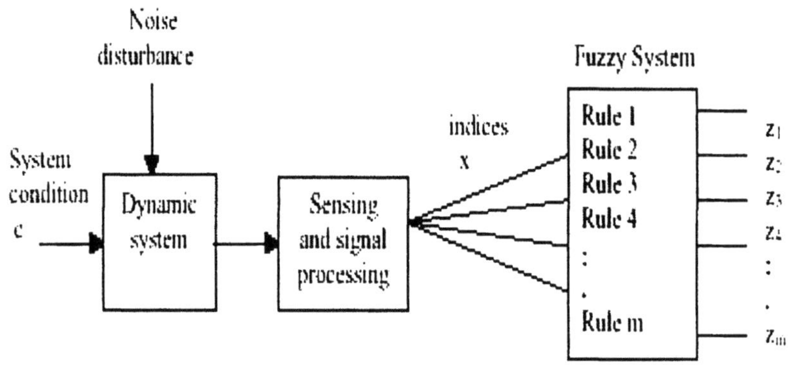

Figure 8.2. Fuzzy System for Plant Monitoring and Diagnosis.

8.2 Fractal Dimension of a Geometrical Object

Fractal geometry is a mathematical tool for dealing with complex systems that have no characteristic length scale. A well-known example is the shape of a coastline (Semmes, 2000). When we see two pictures of a coastline on two different scales, we cannot tell which scale belongs to which picture: both look the same. This means that the coastline is scale invariant or, equivalently, has no characteristic length scale. Other examples in nature are rivers, cracks, mountains, and clouds. Scale-invariant systems are usually characterized by non-integer ("fractal") dimensions.

The dimension tells us how some property of an object or space changes as we view it at increased detail. There are several different types of dimension. The fractal dimension d_f describes the space filling properties of an object. Three examples of the fractal dimension are the self-similarity dimension, the capacity dimension, and the Hausdorff-Besicovitch dimension (Tricot, 1994). The topological dimension d_T describes how points within an object are connected together. The embedding dimension d_e describes the space in which the object is contained.

The fractal dimensions d_f are useful and important tools to quantify self-similarity and scaling (Mandelbrot, 1997). Essentially, the dimension tells us how many new pieces are resolved as the resolution is increased. The self-similarity dimension can only be applied to geometrical self-similar objects, where the small pieces are exact copies of the whole object (Mandelbrot, 1987). However the capacity dimension can be used to analyze irregularly shaped objects that are statistically self-similar. On the other hand, the Hausdorff-Besicovitch dimension requires more complex mathematical tools (Pesin, 1999). For this reason, we will limit our discussion here to the capacity dimension.

A ball is the set of points within radius r of a given point. We determine N(r) the minimum number of balls required so that each point in the object is contained within at least one ball of radius r. In order to cover all the points of the object, the balls may need to overlap. The capacity dimension defined as above is a measure of the space filling properties of an object because it gives us an idea of how much work is needed to cover the object with balls of changing size.

A useful method to determine the capacity dimension is to choose balls that are the non-overlapping boxes of a rectangular coordinate grid. N(r) is then the number of boxes with side of length r that contain at least one point of the object. Efficient algorithms have been developed to perform this "box counting" for different values of r, and thus determine the box counting dimension as the best fit of log N(r) versus log (1/r).

The fractal dimension d_f characterizes the space-filling properties of an object. The topological dimension d_T characterizes how the points that make up the object are connected together. It can have only integer values. Consider a line that is so long and wiggly that it touches every point in a plane and thus covers an area. Because it covers a plane, its space-filling fractal dimension $d_f = 2$. However, no matter how wiggly it is, it is still a line and thus has topological dimension $d_T=1$. Thus, the essence of a fractal is that its space-filling properties are larger than one anticipates from its topological dimension. Thus we can now present a formal definition of a fractal (Mandelbrot, 1987), namely, that an object is a fractal if and only if $d_f > d_T$. However, there is no one definition that includes all the objects or processes that have fractal properties.

Despite the identification of fractals in nearly every branch of science, too frequently the recognition of fractal structure is not accompanied with any additional insight as to its cause. Often we do not even have the foggiest idea as to the underlying dynamics leading to the fractal structure. The chaotic dynamics of non-linear systems, on the other hand, is one area where considerable progress has been made in understanding the connection with fractal geometry. Indeed, chaotic dynamics and fractal geometry have such a close relationship that one of the hallmarks of chaotic behavior has been the manifestation of fractal geometry, particularly for strange attractors in dissipative systems (Moon, 1992). For a practical definition we take a "strange attractor", for a dynamic system, to be an attracting set with fractal dimension. For example, the famous Lorenz strange attractor has a fractal dimension of about 2.06. Also, we think that beyond only this relationship between strange attractors and the fractal dimension of the set, there is a deeper relationship between the underlying dynamics of a system and the fractal nature of its behavior.

Recently, considerable progress has been made in understanding the complexity of an object through the application of fractal concepts and dynamic scaling theory (Peters, 1994). For example, financial time series show scaled properties suggesting a fractal structure (Castillo & Melin, 2001). The fractal dimension of a geometrical object can be defined as follows:

$$d = \lim_{r \to 0} [\ln N(r)] / [\ln(1/r)] \tag{8.3}$$

where N(r) is the number of boxes covering the object and r is the size of the box. An approximation to the fractal dimension can be obtained by counting the number of boxes covering the boundary of the object for different r sizes and then performing a logarithmic regression to obtain d (box counting algorithm). In Figure 8.3, we illustrate the box counting algorithm for a hypothetical curve C. Counting the number of boxes for different sizes of r and performing a logarithmic linear regression, we can estimate the box dimension of a geometrical object with the following equation:

$$\ln N(r) = \ln\beta - d \ln r \tag{8.4}$$

this algorithm is illustrated in Figure 8.4.

The fractal dimension can be used to characterize an arbitrary object. The reason for this is that the fractal dimension measures the geometrical complexity of objects. In this case, a time series can be classified by using the numeric value of the fractal dimension (d is between 1 and 2 because we are on the plane xy). The reasoning behind this classification scheme is that when the boundary is smooth the fractal dimension of the object will be close to one. On the other hand, when the boundary is rougher the dimension will be close to a value of two.

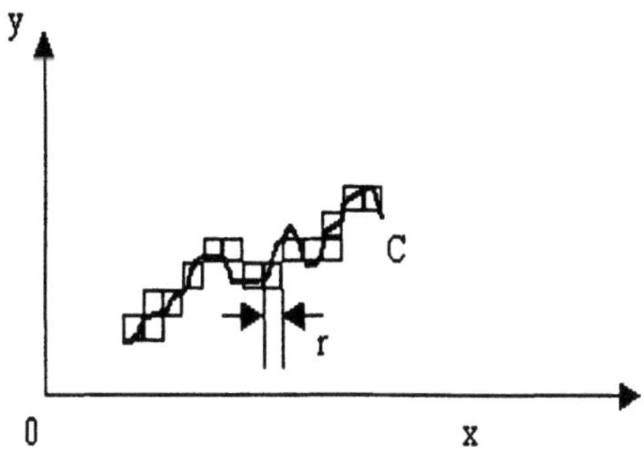

Figure 8.3. Box Counting Algorithm for a Curve C.

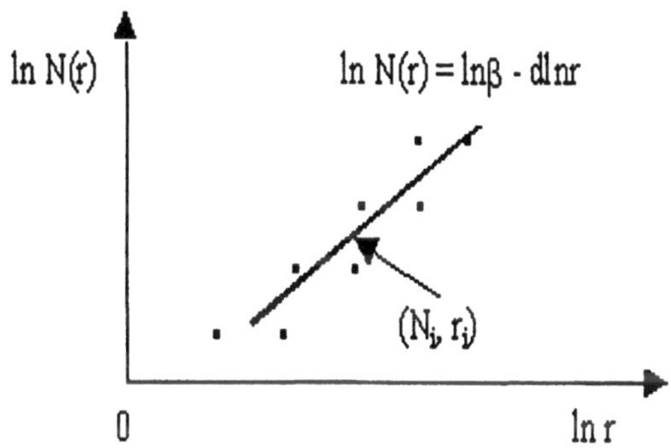

Figure 8.4. Logarithmic Regression to find the Fractal Dimension.

8.3 Fuzzy Estimation of the Fractal Dimension

The fractal dimension of a geometrical object is a crisp numerical value measuring the geometrical complexity of the object (Melin & Castillo, 2001). However, in practice it is difficult to assign a unique numerical value to an object. It is more appropriate to assign a range of numerical values in which there exists a membership degree for this object. For this reason, we will assign to an object O a fuzzy set with membership function μ_o, which measures the membership degree for that object.

Lets consider that the object O is in the plane xy, then a suitable membership function is a generalized bell function:

$$\mu_o(x) = 1 / [\, 1 + |\,(x-c) / a\,|^{2b}\,] \tag{8.5}$$

where a, b and c are the parameters of the membership function. Of course other types of membership functions could be used depending on the characteristics of the application. The important thing here is that the membership functions need to be defined over the appropriate universe of discourse X, and the parameters need to be adjusted to give the appropriate linguistic interpretation.

By using the concept of a fuzzy set (Zadeh, 1965, 1975) we are in fact generalizing the mathematical concept of the fractal dimension. In fact, our definition of the fuzzy fractal dimension is as follows.

Definition 8.1: Let O be an arbitrary geometrical object in the plane xy. Then the fuzzy fractal dimension is the pair: (d_o, μ_o) where d_o is the numerical value of the fractal dimension calculated by the box counting algorithm, and μ_o is the membership function for the object.

With this new definition we can account for the uncertainty in the estimation of the fractal dimension of an object. Also, this new definition enables easier pattern recognition for objects, because it is not necessary to match an exact numerical value to recognize a particular object (Bezdek, 1981), (Yager, 1994). Of course, for a particular application the values of the membership function parameters a, b, c will need to be adjusted with an iterative method, like with neural networks (Omidvar & Elliot, 1997) or genetic algorithms (Mitchell, 1998).

8.4 Plant Monitoring with Fuzzy-Fractal Approach

In this section, we show how to implement a fuzzy rule-based expert monitoring system with two basic sensors: temperature, and pressure. We also use as input the fuzzy fractal dimension of the time series of the measured variables. Individual sensors can identify three linguistic values (normal, high, and low). The three inputs can be combined to give 9 different scenarios. With the perfectly normal case (where all three input variables have normal values), there are additionally 8 more cases where combinations of abnormal readings can be observed.

Let x_1 be the temperature, x_2 the pressure, x_3 the fuzzy fractal dimension, and y the diagnostic statement. Let L_i, N_i, and H_i, represent the three sets of low range, normal range, and high range for input data x_i, where i = 1, 2, or 3. Furthermore, let C_1, C_2,..., C_9 be the individual scenarios that could happen for each combination of the different data sets. The fuzzy rules have the general form:

$$R^{(0)}: \text{IF } x_1 \text{ is } N_1 \text{ AND } x_2 \text{ is } N_2 \text{ AND } x_3 \text{ is } N_3 \text{ THEN } y \text{ is } C_1 \quad (8.6)$$

... ...

$$R^{(i)}: \text{IF } x_1 \text{ is } V_1 \text{ AND } x_2 \text{ is } V_2 \text{ AND } x_3 \text{ is } V_3 \text{ THEN } y \text{ is } C_i$$

... ...

$$R^{(26)}: \text{IF } x_1 \text{ is } H_1 \text{ AND } x_2 \text{ is } H_2 \text{ AND } x_3 \text{ is } H_3 \text{ THEN } y \text{ is } C_9$$

In this case, V_i represents L_i, H_i, or N_i, depending on the condition for the plant. Experts have to provide their knowledge in plant monitoring to label the individual cases Ci for i = 1, 2,..., 9. Also, the membership functions for the linguistic values of variables have to be defined according to historical data of the problem and expert knowledge.

We can use the Fuzzy Logic Toolbox of the MATLAB programming language to implement the fuzzy monitoring system described above. In this case, we need to specify the particular fuzzy rules and the corresponding membership functions for the problem (Yen & Langari, 1999). We show below a sample implementation of a condition monitoring system using the MATLAB language. First, we show in Figure 8.5 the general architecture of the fuzzy monitoring system. In this figure, we can see the input linguistic variables (temperature, pressure, and fractal dimension) and the output linguistic variable (condition of the plant) of the fuzzy monitoring system. Of course, in this case the fractal dimension is estimated using the box counting algorithm, which was implemented also as a computer program in MATLAB.

In Figure 8.6 the implementation of the fuzzy rule base in MATLAB is shown. The actual 27 rules were defined according to expert knowledge on the process. In Figure 8.7, the Gaussian membership functions for the output variable (condition of the plant) are shown. In Figure 8.8 the non-linear surface for the problem of plant monitoring is shown. Finally, the fuzzy rule viewer of MATLAB is shown in Figure 8.9. In this case, the fuzzy rule viewer is used to calculate the output condition of the plant for specific values of the input variables. We have to point out that a Mamdani fuzzy inference system with max-min operators was used in the implementation of this fuzzy system.

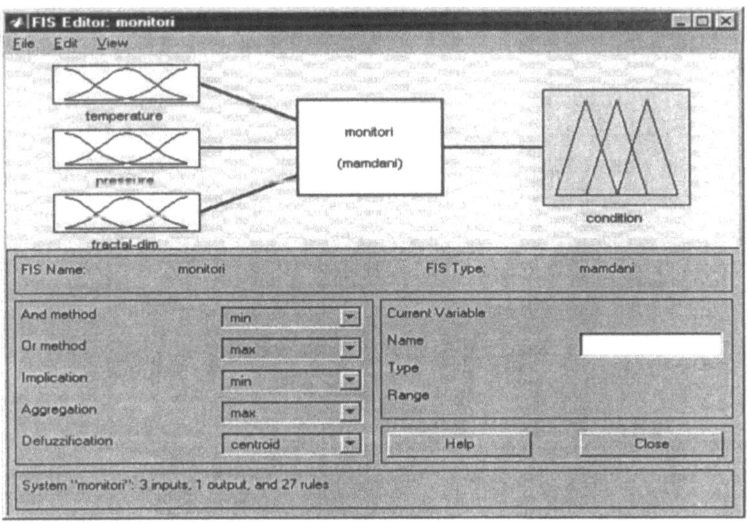

Figure 8.5. Architecture of the fuzzy system for plant monitoring.

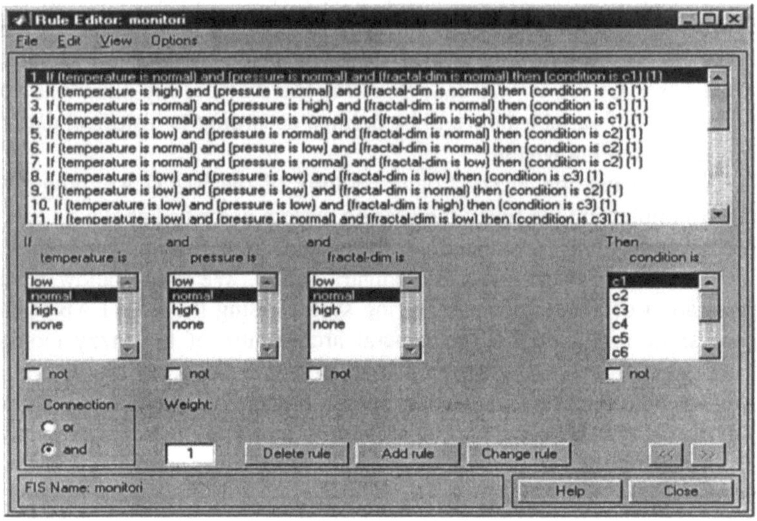

Figure 8.6. Fuzzy rule base for plant monitoring.

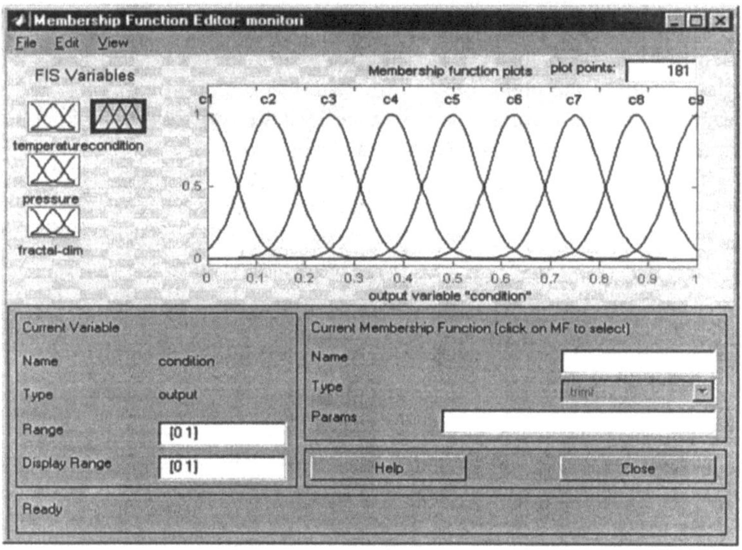

Figure 8.7. Membership functions of the output variable.

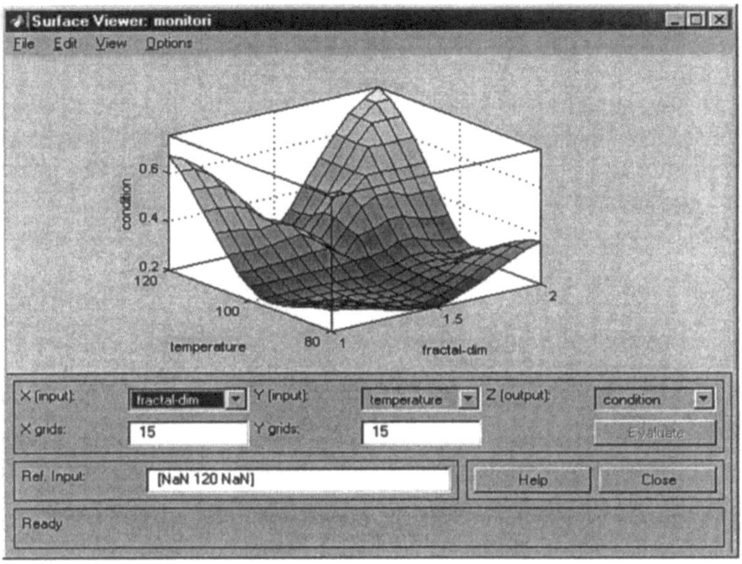

Figure 8.8. Non-linear surface for plant monitoring.

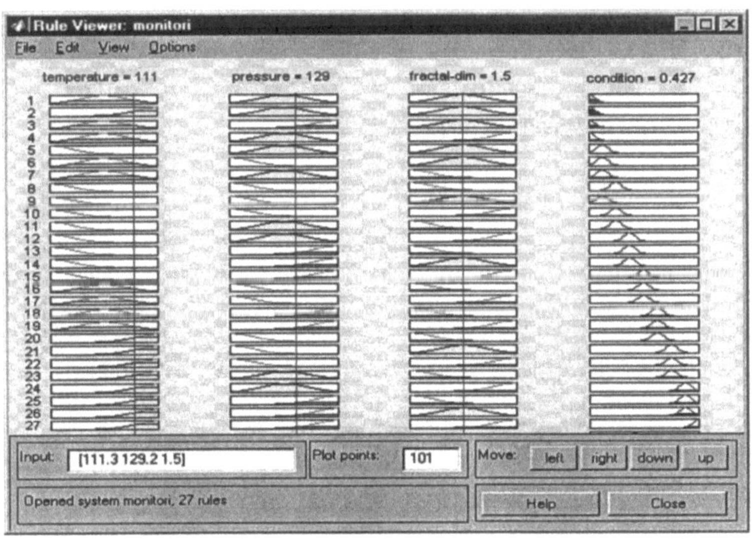

Figure 8.9. Fuzzy rule viewer for specific values of the input variables.

8.5 Experimental Results

In this section, numerical simulations are given to demonstrate the workability of the 27 rules given in the previous section representing a monitoring expert system.
Case 1. In this case, the first rule for normal condition is tested. The temperature profile $x_1(t)$, and pressure profile $x_2(t)$ are generated according to the following equations:

$$x_1(t) = 100 + 4\sin(\pi t/20)e(-t/20) + n(t) \tag{8.7}$$

$$x_2(t) = 120 + 40\sin(\pi t/20)e(-(t - 20)/20) + n(t) \tag{8.8}$$

Where the function n(t) is random white noise and e(t) the exponential function of the form:

$$e(t) = e^t \text{ if } 0 < t < 100,\ e(t) = 0 \text{ otherwise.} \tag{8.9}$$

The $x_3(t)$ variable is calculated from the time series of the $x_1(t)$ and $x_2(t)$ variables using the fractal dimension concept. The above formulas create three separate time intervals, $I_1 = [7, 14]$, $I_2 = [27, 34]$, and $I_3 = [47, 54]$. These data profiles will result in the first rule being valid in I_1, the second rule being valid in I_2, and the third rule being valid in I_3. Thus, the corresponding membership function should yield the value 1 for $t \in I_1$ and the value 0 for everywhere else. Similarly, for the other two intervals.
Case 2. In this case, the rule $R^{(20)}$ is tested. The temperature profile $x_1(t)$, and pressure profile $x_2(t)$ are generated according to the following equations:

$$x_1(t) = 100 + 4\sin(\pi t/40)e(-t/20) + n(t) \tag{8.10}$$

$$x_2(t) = 120 - 40\sin(\pi t/40)e(-t/20) + n(t) \tag{8.11}$$

Where the function n(t) is random white noise and e(t) the exponential function described in the previous case. These formulas create an interval $I_4 = [10, 30]$ where the rule $R^{(20)}$ will be valid, i.e., the corresponding membership function should yield the value 1for $t \in I_4$ and the value 0 for everywhere else.
Case 3. In this case, the rule $R^{(26)}$ is tested. The temperature profile $x_1(t)$, and pressure profile $x_2(t)$ are generated according to the following equations:

$$x_1(t) = 100 + 4\sin(\pi t/40)e(-t/20) + n(t) \tag{8.12}$$

$$x_2(t) = 120 + 40\sin(\pi t/40)e(-t/20) + n(t) \tag{8.13}$$

Where the function n(t) is random white noise and e(t) the exponential function described in case 1. These formulas create an interval $I_4 = [10, 30]$ where the rule

$R^{(26)}$ will be valid, i.e., the corresponding membership function should yield the value 1 for $t \in I_5$ and the value 0 for everywhere else.

Figures 8.10 and 8.11 show the plot of the data $x_1(t)$ for temperature, and $x_2(t)$ for pressure for Case 1. In the first 20 seconds, $x_2(t)$ is in the normal range, and $x_1(t)$ is gradually moving toward the high range, causing the membership function of condition C_2 to rise toward 1. Similarly, during the time interval [20,40], $x_1(t)$ is in the normal range, and $x_2(t)$ is gradually moving toward the high range, causing the membership function of condition C_3 to rise toward 1; and during the time interval [40,60], $x_1(t)$ and $x_2(t)$ are in the normal range, causing the membership function of condition C1 to rise toward 1.

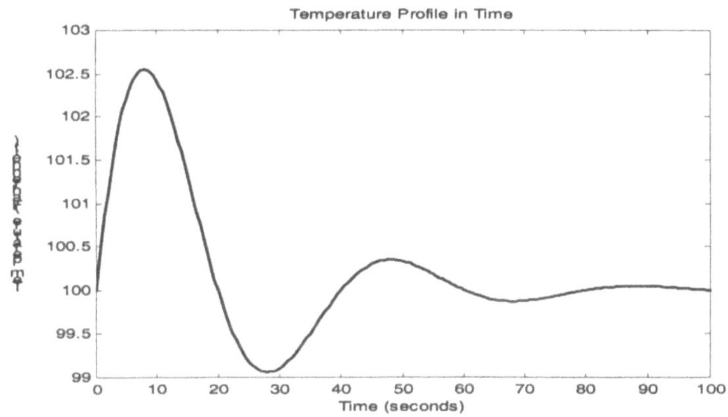

Figure 8.10. Temperature profile in time for case 1.

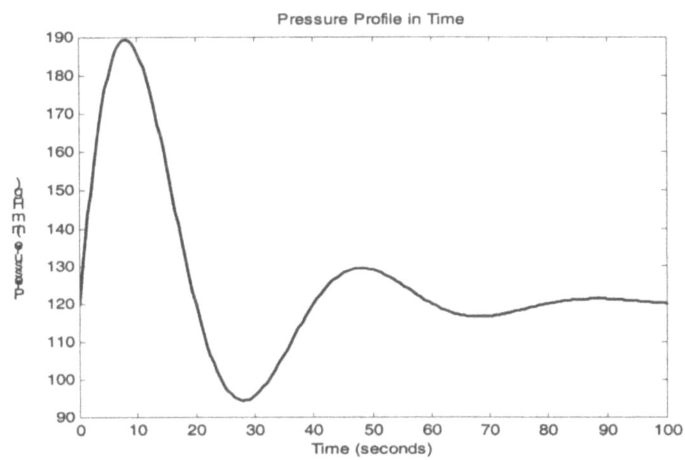

Figure 8.11. Pressure profile in time for case 1.

Based on the examples presented in this section, we see that using fuzzy logic in monitoring and diagnostics always results in improved performance. Also, the use of the fractal dimension improves the accuracy of the method. We have compared the success rate of the fuzzy-fractal approach against the use of only fuzzy logic, using the simulated data described before. The results are shown in Table 8.1. We are using in all of the cases a specific biochemical reactor. The reactor is considered to be in a different condition in each of the three cases. The comparison is between the results of the intelligent system using the fuzzy-fractal approach, and a computer program using only fuzzy logic with the Mamdani approach.

Table 8.1. Success Rate of the fuzzy-fractal approach for monitoring.

Applications	Fuzzy-Fractal	Fuzzy Logic
condition monitoring in a biochemical reactor (case 1)	98%	82%
condition monitoring in a biochemical reactor (case 2)	86%	73%
condition monitoring in a biochemical reactor (case 3)	90%	79%

It should be pointed out that no matter what techniques are used, there is no guarantee of success because monitoring and diagnosis is a process of abduction. First, the training samples may not represent all the patterns of different system conditions. An effective solution to this problem is to add more training samples. Second the patterns of different system conditions overlap and/or are inseparable owing to the definition of system conditions and the use of monitoring indices.

Finally, it is interesting to compare the performance of the fuzzy-fractal approach with that of using only fuzzy logic (see Table 8.1). We see that the fuzzy-fractal approach outperforms the fuzzy logic approach by at least 10% in all the cases. This demonstrates that the hybrid fuzzy-fractal approach is indeed a more effective method of monitoring and diagnostics and, in general outperforms the use of fuzzy logic alone. Another way of saying this, is that the use of the fractal dimension improves the performance of the fuzzy system for problems of monitoring and diagnostics.

8.6 Summary

In this paper a hybrid fuzzy-fractal approach for plant monitoring has been proposed. An implementation in the MATLAB programming language has been shown, to describe in more detail the advantages of the new approach. The hybrid fuzzy-fractal approach combines the advantages of fuzzy logic (expert knowledge representation) with the advantages of the fractal dimension concept (ability to measure object complexity), to achieve efficient monitoring and diagnostics. Simulation results are shown to validate the fuzzy-fractal approach. We have compared the performance of the fuzzy-fractal approach against the use of fuzzy logic, the results show that the use of the fractal dimension improves the performance approximately 10%. A problem yet to be considered, is how to automatically learn (or adapt) the membership functions and rules of the fuzzy system using real data for the problem. A genetic approach could be used to evolve the fuzzy system (including the fractal dimension).

Chapter 9

Adaptive Control of Non-Linear Plants

We describe in this chapter adaptive model-based control of non-linear plants using soft computing techniques. First, the general concept of adaptive model-based control is described. Second, the use of fuzzy logic for adaptive control is described. Third, a neuro-fuzzy approach is proposed to learn the parameters of the fuzzy system for control. A specific non-linear plant is used to test the hybrid approach for adaptive control. A particular stepping motor was used as test bed in the experiments. The results of the neuro-fuzzy approach were good, both in accuracy and efficiency.

Adaptive control is a method of designing a controller with some adjustable parameters and an embedded mechanism for adjusting these parameters (Castillo & Melin, 2001). Adaptive controllers have been used mainly to improve the controller's performance online. For each control cycle, the adaptive algorithm is normally implemented in three basics steps:

(1) Observable data are collected to calculate the controller's performance,
(2) The controller's performance is used as a guidance to calculate the adjustment to a set of controller parameters,
(3) The controller's parameters are then adjusted to improve the performance of the controller in the next cycle.

Normally, an adaptive controller is designed based on one of the available techniques. Each technique is originally designed for a specific class of dynamic system. The controller is then adjusted as data are collected during run time to extend its effectiveness to control a large class of dynamic systems.

A typical application of adaptive control was to calibrate a system at startup. In this case, a controller is also designed for a specific class of dynamic systems. However the parameters that characterize the dynamic behavior of a particular system might not be known in advance. A controller is then designed and arbitrary values are assigned to initialize these parameters. After a few control cycles, parameters are adjusted to converge to the actual parameters of the system.

This approach is often used for cases in which a system is designed to handle a variable payload. The payload is different each time; e.g., a crane is used to pick up a sizeable object. The payload will alter the basic dynamic behavior of a dynamic system. Adaptive control is normally used to calibrate these parameters that characterize such dynamic behavior.

Traditionally, there are four basic approaches for adaptive control: (i) gain scheduling, (ii) model reference adaptive system, (iii) self-tuning regulator, and (iv) dual control. Gain scheduling is a method of adjusting the control signal based on a known look-up table describing changes of a dynamic system. The model reference adaptive system is a method of comparing the performance of the actual system against an assumed mathematical model that describes the actual system, and designing control input to drive this comparison error to zero. Self-tuning regulator is a method of updating the parameters of a model that describes the plant based on observed data, and channeling the updated information into the controller that is designed based on these parameters. Dual control is a method of extending adaptive control to stochastic model dealing with uncertainties.

In this chapter, an extension to the principle of adaptive control using soft computing techniques is proposed. First, the concept of adaptive fuzzy control is described. A neuro-fuzzy approach is used to learn the parameters of the fuzzy system using real data from the plant (Chen & Pham, 2001). The proposed adaptive control was tested with a specific non-linear plant, to evaluate the performance of this approach. A stepping motor was used as the non-linear plant in this study. The simulation results clearly show the advantage of using soft computing techniques to improve the concept of adaptive control.

9.1 Fundamental Adaptive Fuzzy Control Concept

Adaptive fuzzy control is an extension of fuzzy control theory to allow the fuzzy controller, extending its applicability, either to a wider class of uncertain systems or to fine-tune the parameters of a system to accuracy (Castillo & Melin, 2001). In this scheme, a fuzzy controller is designed based on knowledge of a dynamic system. This fuzzy controller is characterized by a set of parameters. These parameters are either the controller constants or functions of a model's constants.

A controller is designed based on an assumed mathematical model representing a real system. It must be understood that the mathematical model does not completely match the real system to be controlled. Rather, the mathematical model is seen as an approximation of the real system. A controller designed based on this model is assumed to work effectively with the real system if the error between the actual system and its mathematical representation is relatively insignificant. However, there exists a threshold constant that sets a boundary for the effectiveness of a controller. An error above this threshold will render the controller ineffective toward the real system.

An adaptive controller is set up to take advantage of additional data collected at run time for better effectiveness. At run time, data are collected periodically at the beginning of each constant time interval, $t_n = t_{n-1} + \Delta t$, where Δt

is a constant measurement of time, and $[t_n, t_{n-1})$ is a duration between data collection. Let D_n be a set of data collected at time $t = t_n$. It is assumed that at any particular time, $t = t_n$, a history of data $\{D_0, D_1, \ldots, D_n\}$ is always available. The more data available, more accurate the approximation of the system will become.

At run time, the control input is fed into both the real system and the mathematical model representing the system. The output of the real system and the output of that mathematical model are collected and an error representing the difference between these two outputs are calculated. Let x(t) be the output of the real system, and y(t) the output of the mathematical model. The error ε(t) is defined as:

$$\varepsilon(t) = x(t) - y(t). \tag{9.1}$$

Figure 9.1 depicts this tracking of the difference between the mathematical model and the real dynamic system it represents.

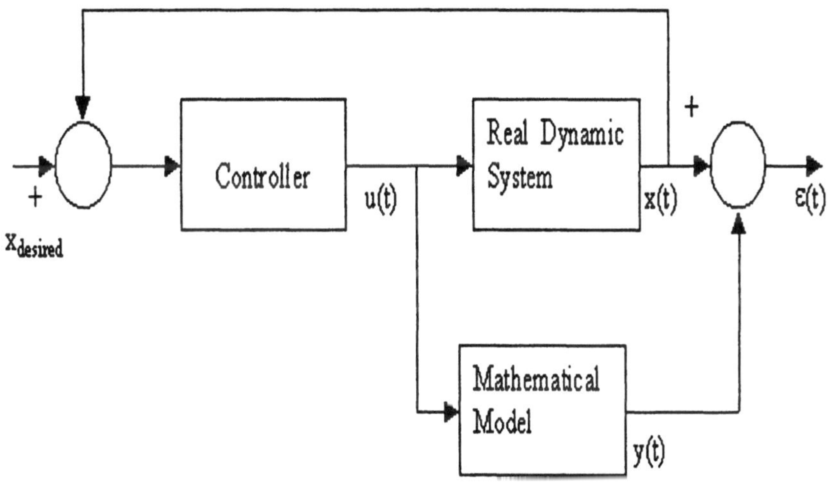

Figure 9.1 Tracking the error function between outputs of a real system and mathematical model

An adaptive controller will be adjusted based on the error function ε(t). This calculated data will be fed into either the mathematical model or the controller for adjustment. Since the error function ε(t) is available only at run time, an adjusting mechanism must be designed to accept this error as it becomes available, i.e., it must evolve with the accumulation of data in time. At any time, $t = t_n$, the set of calculated data in the form of a time series $\{\varepsilon(t_0), \varepsilon(t_1), \ldots, \varepsilon(t_n)\}$ is available and must be used by the adjusting mechanism to update appropriate parameters.

In normal practice, instead of doing re-calculation based on a lengthy set of data, the adjusting algorithm is reformulated to be based on two entities: (i) sufficient information, and (ii) newly collected data. The sufficient information is a numerical variable representing the set of data $\{\varepsilon(t_0), \varepsilon(t_1),..., \varepsilon(t_{n-1})\}$ collected from the initial time t_0 to the previous collecting cycle starting at time $t = t_{n-1}$. The new datum $\varepsilon(t_n)$ is collected in the current cycle starting at time $t = t_n$.

An adaptive controller will operate as follows. The controller is initially designed as a function of a parameter set and state variables of a mathematical model. The parameters can be updated any time during operation and the controller will adjust itself to the newly updated parameters. The time frame is usually divided into a series of equally spaced intervals $\{[t_n,t_{n+1})| \ n = 0,1,2,...; t_{n+1} = t_n + \Delta t\}$. At the beginning of each time interval $[t_n,t_{n+1})$ observable data are collected and the error function $\varepsilon(t_n)$ is calculated. This error is used to calculate the adjustment in the parameters of the controller. New control input $u(t_n)$ for the time interval $[t_n,t_{n+1})$ is then calculated based on the newly calculated parameters and fed into both the real dynamic system under control and the mathematical model upon which the controller is designed. This completes one control cycle. The next control cycle will consist of the same steps repeated for the next time interval $[t_{n+1},t_{n+2})$, and so on.

A dynamic system can be described by a set of fuzzy if-then rules that correlate the input and the output. These rules include a set of parameters that is used to uniquely calculate the estimated output of a system given the inputs and current states of the system.

A dynamic system is mathematically modeled as a set of fuzzy if-then rules:

R^1: IF $x_1(t_n)$ is X_{11} AND ...AND $x_i(t_n)$ is X_{1i}
THEN $y_1(t_{n+1}) = a_{11}x_1(t_n) + a_{12}x_2(t_n) + \ldots + a_{1i}x_i(t_n) + b_{11}u_1(t_n) + b_{12}u_2(t_n) + \ldots + b_{1i}u_j(t_n)$

R^2: IF $x_1(t_n)$ is X_{21} AND ...AND $x_i(t_n)$ is X_{2i}
THEN $y_2(t_{n+1}) = a_{21}x_1(t_n) + a_{22}x_2(t_n) + \ldots + a_{2i}x_i(t_n) + b_{21}u_1(t_n) + b_{22}u_2(t_n) + \ldots + b_{2i}u_j(t_n)$

... ...

R^k: IF $x_1(t_n)$ is X_{k1} AND ...AND $x_i(t_n)$ is X_{ki}
THEN $y_k(t_{n+1}) = a_{k1}x_1(t_n) + a_{k2}x_2(t_n) + \ldots + a_{ki}x_i(t_n) + b_{k1}u_1(t_n) + b_{k2}u_2(t_n) + \ldots + b_{ki}u_j(t_n)$

where $x_1, x_2, \ldots, x_i$ are the observable state variables of the real system, $y_1, y_2, \ldots, y_k$ are the calculated state variables of the mathematical model, and $u_1, u_2, \ldots, u_j$ the control inputs to both the real system and the mathematical model.

The above fuzzy mathematical model is characterized by a set of parameters $\{a_{ki}, b_{kj}\}$ These parameters will determine the behavior of the mathematical model. Theoretically, the model will approximate the real system. The more accurately the model approximates the real dynamic system, the better the controller designed based on this model behaves.

Similarly, a fuzzy controller, if designed based on the fuzzy mathematical model above, will have the following form of fuzzy rules:

R^1: IF $x_1(t_n)$ is X_{11} AND ...AND $x_i(t_n)$ is X_{1i}
THEN $u_1(t_n) = k_{11}x_1(t_n) + k_{12}x_2(t_n) + \ldots + k_{1i}x_i(t_n)$
R^2: IF $x_1(t_n)$ is X_{21} AND ...AND $x_i(t_n)$ is X_{2i}
THEN $u_2(t_n) = k_{21}x_1(t_n) + k_{22}x_2(t_n) + \ldots + k_{2i}x_i(t_n)$
... ...
R^j: IF $x_1(t_n)$ is X_{k1} AND ...AND $x_i(t_n)$ is X_{ki}
THEN $u_j(t_n) = k_{j1}x_1(t_n) + k_{j2}x_2(t_n) + \ldots + k_{ji}x_i(t_n)$

In this case, the set of parameters $\{k_{ji}\}$ characterizes the controller. It can be understood intuitively that the parameter set $\{k_{ji}\}$ is designed as a function of the parameter set $\{a_{ki}, b_{kj}\}$, so that the state variables of the fuzzy model are driven to a target point with stability.

Therefore, it is natural for designing the controller to either adjust the system parameters $\{a_{ki}, b_{kj}\}$, which will lead to the adjustment of the control parameters $\{k_{ji}\}$, or directly adjust the controller parameters $\{k_{ji}\}$ as a scheme to implement adaptive strategies. The adjustment of the system parameters is sometimes called system identification. This system identification process is often applied to a system that can be described by a time-varying parameter set in the fuzzy model.

Adjusting the system parameters is normally carried out to improve the performance of the overall closed-loop system, or to improve accuracy of the assumed model upon which the controller is design. The process of calculating the adjustment to the parameters is the adjusting mechanism mentioned as part of the adaptive control problem.

9.2 Basic Concepts of Stepping Motors

Stepping motors come in two varieties, *permanent magnet* and *variable reluctance* (there are also *hybrid* motors, which are indistinguishable from permanent magnet motors from the controller's point of view). Lacking a label on the motor, you can generally tell the two apart by feel when no power is applied. Permanent magnet motors tend to "cog" as you twist the rotor with your fingers, while variable reluctance motors almost spin freely (although they may cog slightly because of residual magnetization in the rotor). You can also distinguish between the two varieties with an ohmmeter. Variable reluctance motors usually have three (sometimes four) windings, with a common return, while permanent magnet motors usually have two independent windings, with or without center taps. Center-tapped windings are used in unipolar permanent magnet motors (Kenjo, & Sugawara, 1994).

Stepping motors come in a wide range of angular resolution. The coarsest motors typically turn 90 degrees per step, while high-resolution permanent magnet motors are commonly able to handle 1.8 or even 0.72 degrees per step. With an appropriate controller, most permanent magnet and hybrid motors can be run in half-steps, and some controllers can handle smaller fractional steps or micro-steps. For both, permanent magnet and variable reluctance stepping motors, just one winding of the motor is energized, the rotor (under no load) will snap to a fixed angle and then hold that angle until the torque exceeds the holding torque of the motor, at which point, the rotor will turn, trying to hold at each successive equilibrium point.

9.2.1 Variable Reluctance Motors

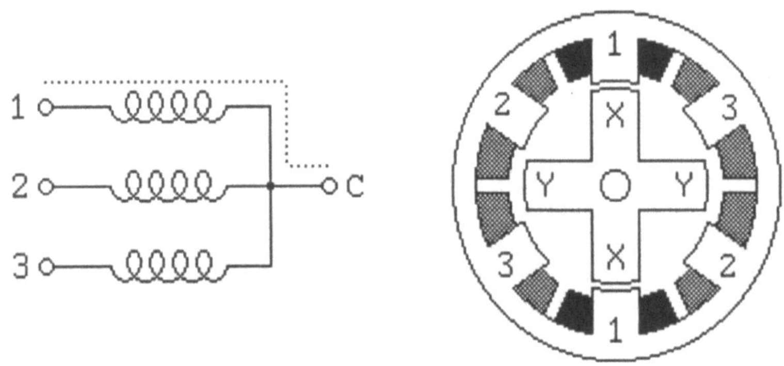

Figure 9.2 Example of a variable reluctance motor.

If your motor has three windings, typically connected as shown in the schematic diagram in Figure 9.2, with one terminal common to all windings, it is most likely a variable reluctance stepping motor. In use, the common wire typically goes to the positive supply and the windings are energized in sequence. The cross section shown in Figure 9.2 is of 30 degree per step variable reluctance motor. The rotor in this motor has 4 teeth and the stator has 6 poles, with each winding wrapped around two opposite poles. With winding number 1 energized, the rotor teeth marked X are attracted to this winding's poles. If the current through winding 1 is turned off and winding 2 is turned on, the rotor will rotate 30 degrees clockwise so that the poles marked Y line up with the poles marked 2.

There are also variable reluctance stepping motors with 4 and 5 windings, requiring 5 or 6 wires. The principle for driving these motors is the same as that for the three winding variety, but it becomes important to work out the correct order to energize the windings to make the motor step nicely. The motor geometry illustrated in Figure 9.2, giving 30 degrees per step, uses the fewest number of rotor teeth and stator poles that performs satisfactorily. Using more motor poles and more rotor teeth allows construction of motors with smaller step angle.

9.2.2 Unipolar Motors

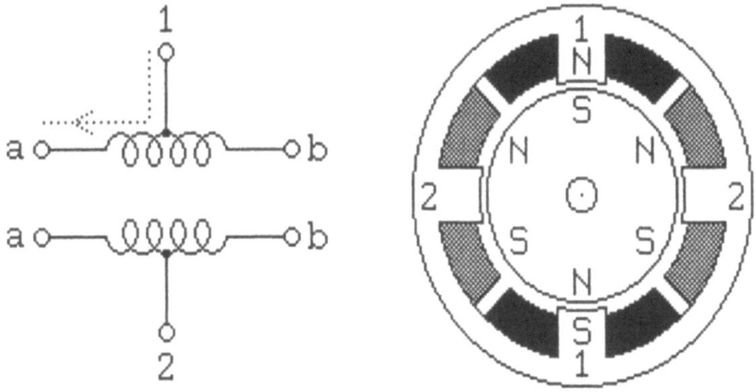

Figure 9.3 Example of a unipolar motor.

Unipolar stepping motors, both Permanent magnet and hybrid stepping motors with 5 or 6 wires are usually wired as shown in the schematic in Figure 9.3, with a center tap on each of two windings. In use, the center taps of the windings are typically wired to the positive supply, and the two ends of each winding are alternately grounded to reverse the direction of the field provided by that winding. The motor cross section shown in Figure 9.3 is of a 30 degree per step permanent magnet or hybrid motor. Motor winding number 1 is distributed between the top and bottom stator pole, while motor winding number 2 is distributed between the left and right motor poles. The rotor is a permanent magnet with 6 poles, 3 south and 3 north, arranged around its circumference.

For higher angular resolutions, the rotor must have proportionally more poles. The 30 degree per step motor in the figure is one of the most common permanent magnet motor designs, although 15 and 7.5 degree per step motors are widely available. Permanent magnet motors with resolutions as good as 1.8 degrees per step are made, and hybrid motors are routinely built with 3.6 and 1.8 degrees per step, with resolutions as fine as 0.72 degrees per step available. As shown in the figure, the current flowing from the center tap of winding 1 to terminal a causes the top stator pole to be a north pole while the bottom stator pole is a south pole. This attracts the rotor into the position shown. If the power to winding 1 is removed and winding 2 is energized, the rotor will turn 30 degrees, or one step. Note that the two halves of each winding are never energized at the same time.

9.2.3 Bipolar Motors

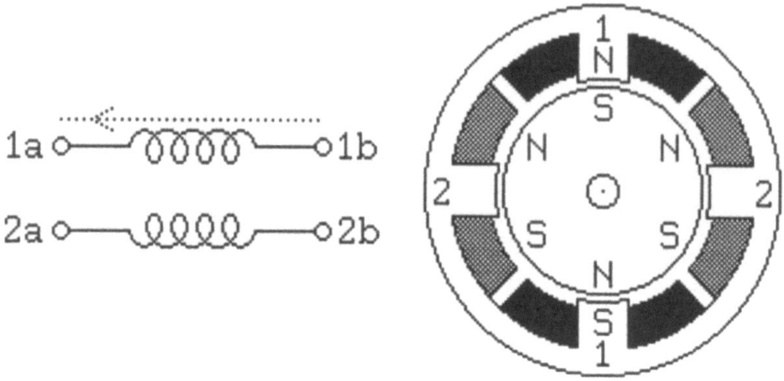

Figure 9.4 Example of a bipolar motor.

Bipolar permanent magnet and hybrid motors are constructed with exactly the same mechanism as is used on unipolar motors, but the two windings are wired more simply, with no center taps. Thus, the motor itself is simpler but the drive circuitry needed to reverse the polarity of each pair of motor poles is more complex. The schematic in Figure 9.4 shows how such a motor is wired, while the motor cross section shown here is exactly the same as the cross section shown in Figure 9.3.

To distinguish a bipolar permanent magnet motor from other 4 wire motors, measure the resistances between the different terminals. It is worth noting that some permanent magnet stepping motors have 4 independent windings, organized as two sets of two. Within each set, if the two windings are wired in series, the result can be used as a high voltage bipolar motor. If they are wired in parallel, the result can be used as a low voltage bipolar motor. If they are wired in series with a center tap, the result can be used as a low voltage unipolar motor.

9.2.4. Dynamics of the Stepping Motor

Each time you step the motor, you electronically move the equilibrium position S radians. This moves the entire curve a distance of S radians, as shown in Figure 9.5:

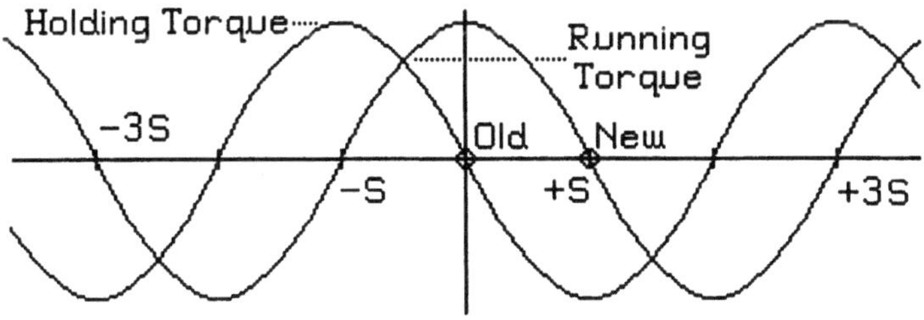

Figure 9.5 Dynamics of a stepping motor.

The first thing to note about the process of taking one step is that the maximum available torque is at a minimum when the rotor is halfway from one step to the next. This minimum determines the *running torque*, the maximum torque the motor can drive as it steps slowly forward. For common two-winding permanent magnet motors with ideal sinusoidal torque versus position curves and holding torque h, this will be $h/(2^{0.5})$. If the motor is stepped by powering two windings at a time, the running torque of an ideal two-winding permanent magnet motor will be the same as the single-winding holding torque.

It should be noted that at higher stepping speeds, the running torque is sometimes defined as the *pull-out torque*. That is, it is the maximum frictional torque the motor can overcome on a rotating load before the load is pulled out of step by the friction. Some motor data sheets define a second torque figure, the *pull-in torque*. This is the maximum frictional torque that the motor can overcome to accelerate a stopped load to synchronous speed. In practice, there is always some friction, so after the equilibrium position moves one step, the rotor is likely to oscillate briefly about the new equilibrium position. The resulting trajectory may resemble the one shown in Figure 9.6:

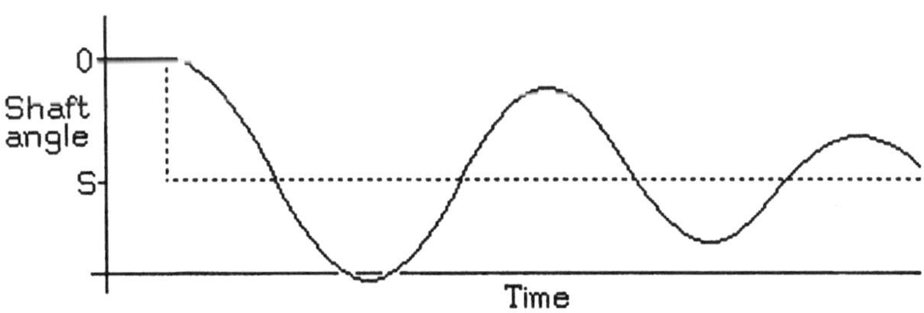

Figure 9.6 Trajectory of the motor rotor.

Here, the trajectory of the equilibrium position is shown as a dotted line, while the solid curve shows the trajectory of the motor rotor. The resonant frequency of the motor rotor depends on the amplitude of the oscillation; but as the amplitude decreases, the resonant frequency rises to a well-defined small-amplitude frequency. This frequency depends on the step angle and on the ratio of the holding torque to the moment of inertia of the rotor. Formally, the small-amplitude resonance can be computed as follows: First, recall Newton's law for angular acceleration:

$$T = \mu A$$

where: T torque applied to rotor, μ moment of inertia of rotor and load, and A angular acceleration, in radians per second per second.

We assume that, for small amplitudes, the torque on the rotor can be approximated as a linear function of the displacement from the equilibrium position. Therefore, Hooke's law applies:

$$T = -k\Theta$$

where: k -- the "spring constant" of the system, in torque units per radian, Θ-- angular position of rotor, in radians.

We can equate the two formulas for the torque to get: $\mu A = -k\Theta$.

Note that acceleration is the second derivative of position with respect to time: $A = d^2\Theta/dt^2$ so we can rewrite this the above in differential equation form:

$$d^2\Theta/dt^2 = -(k/\mu)\,\Theta$$

To solve this, recall that, for: $f(t) = a \sin bt$

The derivatives are:

$$df(t)/dt = ab \cos bt$$
$$d^2f(t)/dt^2 = -ab^2 \sin bt = -b^2 f(t)$$

Note that, throughout this discussion, we assumed that the rotor is resonating. Therefore, it has an equation of motion something like:

$$\Theta = a \sin(2\pi f t)$$

where: a = angular amplitude of resonance , f = resonant frequency .

This is an admissible solution to the above differential equation if we agree that:

$$b = 2\pi f,\ b^2 = k/\mu$$

In practice, this oscillation can cause significant problems when the stepping rate is anywhere near a resonant frequency of the system; the result frequently appears as random and uncontrollable motion.

9.2.5 Control of the Stepping Motor

Stepping motors are widely used in robotics and in the numerical control of machine tools where they have to perform high-precision positioning operations. However, the variations of the mechanical configuration of the drive, which are common to these two applications, can lead to a loss of synchronism for high stepping rates. Moreover, the classical open-loop speed control is weak and a closed-loop control becomes necessary. In this chapter, fuzzy logic is applied to control the speed of a stepping motor drive with feedback. A neuro-fuzzy hybrid approach is used to design the fuzzy rule base of the intelligent system for control. In particular, we used the ANFIS methodology to build a Sugeno fuzzy model for controlling the stepping motor drive. An advanced test bed is used in order to evaluate the tracking properties and the robustness capacities of the fuzzy logic controller.

Stepping motors can be viewed as electric motors without commutators (Kenjo & Sugawara, 1994). Typically, all windings in the motor are part of the stator, and the rotor is either a permanent magnet or, in the case of variable reluctance motors, a toothed block of some magnetically soft material. All of the commutation must be handled externally by the motor controller, and typically, the motors and controllers are designed so that the motor may be held in any fixed position as well as being rotated one way or the other. Most stepping motors can be stepped at audio frequencies, allowing them to spin quite quickly, and with an appropriate controller, they may be started and stopped "on a dime" at controlled orientations.

For some applications, there is a choice between using servomotors and stepping motors. Both types of motors offer similar opportunities for precise positioning, but they differ in a number of ways. Servomotors require analog feedback control systems of some type. Typically, this involves a potentiometer to provide feedback about the rotor position, and some mix of circuitry to drive a current through the motor inversely proportional to the difference between the desired position and the current position. In making a choice between stepping motors and servomotors, a number of issues must be considered; which of these will matter depends on the application. For example, the repeatability of positioning done with a stepping motor depends on the geometry of the motor rotor, while the repeatability of positioning done with a servomotor generally depends on the stability of the potentiometer and other analog components in the feedback circuit.

Stepping motors can be used in simple open-loop control systems; these are generally adequate for systems that operate at low accelerations with static loads, but closed loop control may be essential for high accelerations, particularly if they involve variable loads (Betin, Pinchon & Capolino, 2000). If a stepping motor in an open-loop control system is overtorqued, all knowledge of rotor position is lost and the system must be reinitialized; servomotors are not subject to this problem.

In this chapter, the application of fuzzy logic is proposed to control the speed of a stepping motor drive. The closed-loop control scheme entails in incorporating engineering knowledge into the automatic control system by using the intuition and experience of the designer. This strategy was proposed by Zadeh (1975), to describe complicated systems, which are hard to analyze using traditional mathematics. Indeed, Mamdani (1974) was the first to report on the application of fuzzy logic to control a small laboratory steam engine. The success of this study led many scientists to attempt to control industrial processes such as chemical reactors, automatic trains, or nuclear reactors using fuzzy algorithms. The results of these experiments showed that, fuzzy controllers perform better, or at least as well as, classical controllers. Moreover, this technique offers the advantage of requiring only a simple mathematical model to formulate the algorithm, which can easily be implemented by a digital computer. These features are appreciated for nonlinear processes for which there is no reliable model and complex systems where the model is useless due to the large number of equations involved. Additionally, fuzzy logic is used more frequently for the control of electrical machines such as direct current or induction motors (Li & Lau, 1989). Nevertheless, the main problem with fuzzy logic is that there is no systematic procedure for the design of a fuzzy controller. For this reason, we propose in this paper the use of the ANFIS methodology (Jang, Sun & Mizutani, 1997) to adapt the parameters of the fuzzy system for control. We use this neuro-fuzzy approach to train the controller with real data about the problem.

9.3 Fuzzy Logic Controller of the Stepping Motor

The fuzzy logic controller provides an algorithm, which converts the linguistic control, based on expert knowledge into an automatic control strategy. Therefore, the fuzzy logic algorithm is much closer in spirit to human thinking than traditional logical systems. The main problem with fuzzy logic controller generation is related to the choice of the regulator parameters. For this reason, we apply the ANFIS methodology to adapt the parameters of the fuzzy controller according to real data about the problem. The experiments were made on a system described in Figure 9.7. A program in the PC generates the step input to the system and stores the response. The fuzzy controller is also contained in the PC and acts on the system according to the corresponding responses. We now describe briefly the modules of the system.

1) Motor/Driver: Micro-step motor Vexta PV266-01E with five phases and 500 steps by turn. Power driver Vexta DFR1514A with multi-resolution (Minimum: 500 steps by turn; Maximum: 125000 steps by turn). This specific motor is shown in Figure 9.8.
2) Encoder: Optical encoder Bourns of 40000 steps by turn. This encoder generates two square signals with 90 degree difference. With these signals the magnitude of motor movement is determined. We use the programmable logic device (PLD) of Altera (EP5032) to determine the movement of the motor.

3) Data Acquisition Card: PCL-818 of Advantech with 8 analog inputs and 2 analog outputs (12 bits), 16 digital inputs, 16 digital outputs. The sampling time used is 0.25 ms.
4) Computer/Software: Pentium III with 733 MHz. We design a small real time kernel in C language for control and data acquisition, and the fuzzy controller was programmed in MATLAB (Nakamura, 1997).

The linguistic control rules are established considering the dynamic behavior of the stepping motor drive and analyzing the error and its variation. These control rules are expressed in the following general form:

If Error is LP and Change_Error is LP then Speed = p_1*Error
+q_1*Change_Error + r_1
If Error is LP and Change_Error is MP then Speed = p_2*Error
+q_2*Change_Error + r_2 ...

This is a Sugeno fuzzy model (Sugeno & Kang, 1988) for controlling the stepping motor. We used the ANFIS methodology to estimate the parameters of the membership functions and the consequent functions. We used a fuzzy model of 9 rules and 3 membership functions for each linguistic variable. This was the fuzzy controller that gave the best results.

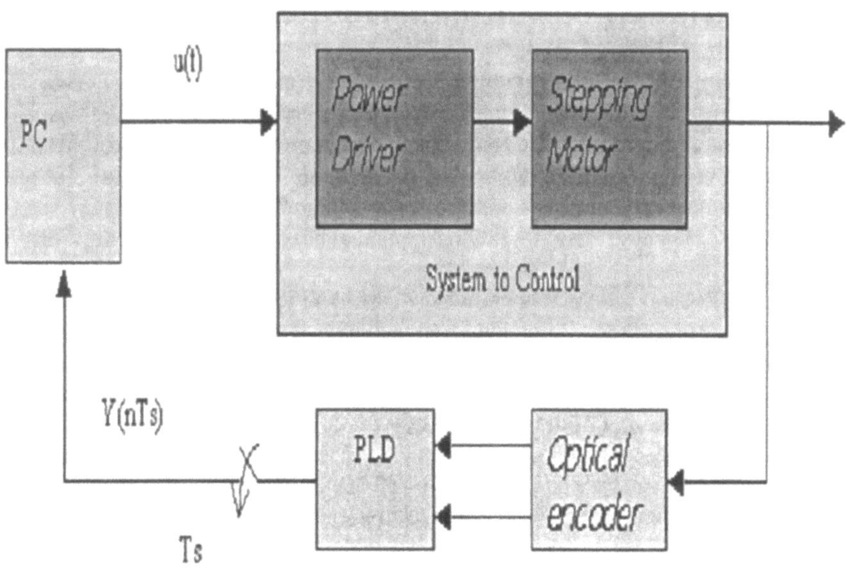

Figure 9.7 Block diagram of the experimental system.

Figure 9.8 Motor Vexta PV266-01E.

9.4 Hardware Implementation of ANFIS

In this section, a brief description of the hardware implementation is given. The ANFIS controller implementation is needed to really achieve the goal of having an intelligent stepping motor. The neuro-fuzzy approach gives to the motor the ability to adapt to changing conditions in the environment. A stepping motor equipped with the intelligent controller will be able to train itself with the new data to update its parameters, and as a consequence be able to change its behavior accordingly.

We have to first mention that the ANFIS implementation required a specific microprocessor to be able to achieve all of the required numerical calculations. The ANFIS methodology requires that training is performed on an adaptive network, using as data a time series of the relevant variables of the problem. For this reason, the "Jstamp" microprocessor was selected. This specific microprocessor uses JAVA as the native programming language, and has 512 Kbytes of RAM memory and 512 Kbytes of FLASH memory. We show in Figure 9.9 the Jstamp microprocessor that was used in our implementation.

Figure 9.9 The Jstamp microprocessor used for the implementation.

9.5 Experimental Results

In this section, the tracking and adaptability features of the fuzzy control applied to the stepping motor are tested using simulation and experimentation. We first show in Figure 9.10 the response of the stepping motor to a sequence of step input signals (we use 400 samples). We show in Figure 9.11 the results of applying the ANFIS methodology with the training data and with the testing data. We used 20 epochs for training and the final error was of 0.000001, which is very good for this application. In Figure 9.12 we plot the predicted values by the fuzzy model and the real values for the system, and the curves are practically indistinguishable. Finally, we show in Figure 9.13 a plot of the difference between both the real and the estimated signal by the fuzzy model. Again, we can note that we have very good results for this application.

We also compared our results with a classical PID controller and with a fuzzy Mamdani controller, to measure how much the adaptive fuzzy approach could improve the performance. Of course, our fuzzy controller (designed with ANFIS) was better in tracking and adaptability than the other controllers. Still, we think that our results could be improved by optimizing the architecture of the controller using genetic algorithms (Jang, Sun & Mizutani, 1997).

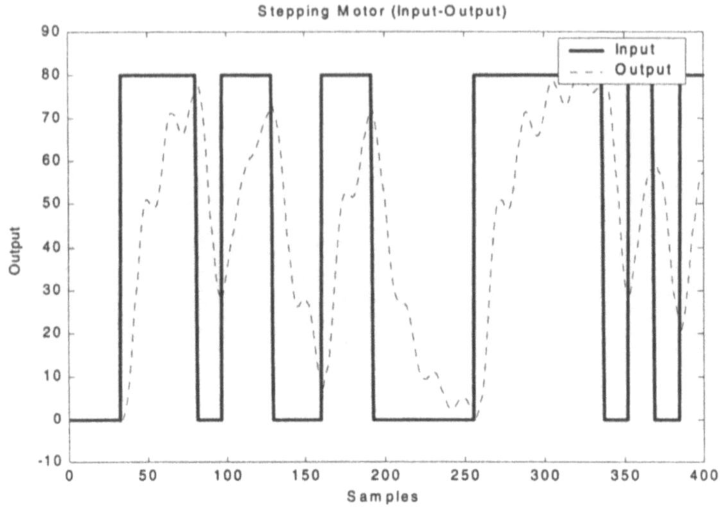

Figure 9.10 Response of the stepping motor to a sequence of step input signals.

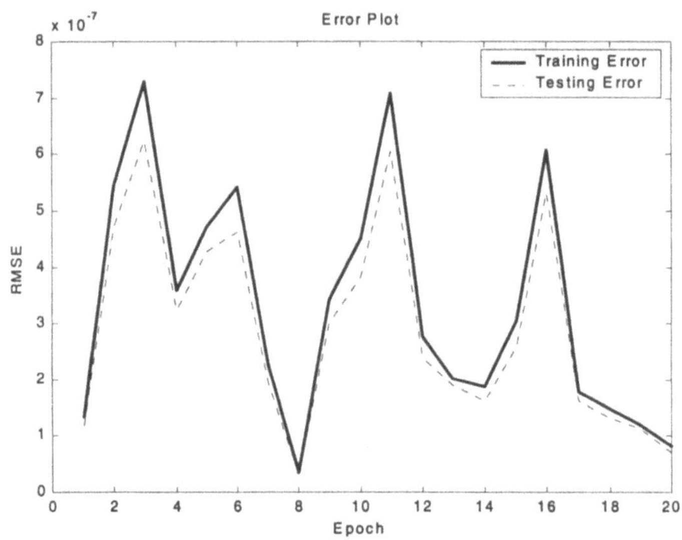

Figure 9.11 Results of applying the ANFIS methodology to the training and testing data.

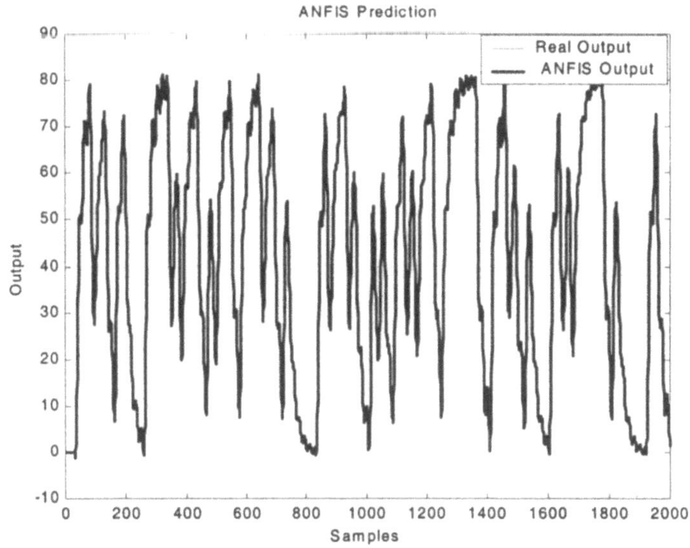

Figure 9.12 Predicted values of the fuzzy model against the real values.

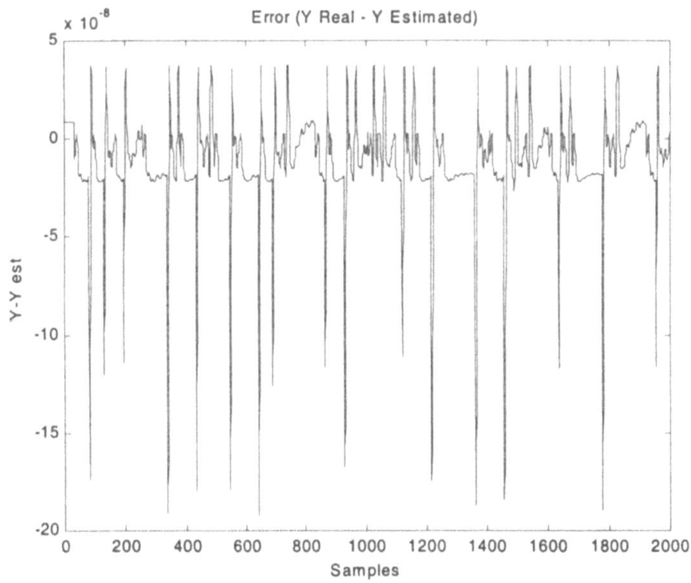

Figure 9.13 Difference between the real and the estimated signal by the fuzzy model.

9.6 Summary

In this chapter, adaptive model-based control of non-linear plants was described. The implementation of an intelligent fuzzy controller for a specific plant was described. A neuro-fuzzy approach was applied to learn the parameters of the fuzzy system for adaptive control. The feasibility of fuzzy control for stepping motor drives has been proved and illustrated by simulation and experimentation. The best parameters for the fuzzy controller were determined by using the ANFIS methodology and also by using simulations of the stepping motor dynamics. An experimental system was used to validate experimentally the tracking ability and the insensibility to plant parameter changes.

The fuzzy controller presented very interesting tracking features and was able to respond to different dynamic conditions. Also, the fuzzy control computation is very inexpensive, and this regulator could be used for the control of machine tools and robotics manipulators (Fu, Gonzalez & Lee, 1997) without significantly increasing the cost of the drive. The only extra cost is for the optical encoder. Another advantage of this method over classical quantitative controllers is that it does not require a fixed sampling time. Therefore, the proposed design confirms the fact that fuzzy control is relevant to the control fast of nonlinear processes such as stepping motor drives where quantitative methods are not always appropriate.

Chapter 10

Automated Quality Control in Sound Speaker Manufacturing

We describe in this chapter the application of a hybrid neuro-fuzzy-fractal approach to the problem of automated quality control in sound speaker manufacturing. Traditional quality control has been done by manually checking the quality of sound after production. This manual checking of the speakers is time consuming and occasionally was the cause of error in quality evaluation. For this reason, we developed an intelligent system for automated quality control in sound speaker manufacturing. The intelligent system has a fuzzy rule base containing the knowledge of human experts in quality control. The parameters of the fuzzy system are tuned by applying the ANFIS methodology using, as training data, a real time series of measured sounds as given by good sound speakers. We also use the fractal dimension to measure the complexity of the sound signal. The intelligent system has been tested in a real plant with very good simulation and experimental results.

10.1 Introduction

We describe in this chapter the application of a neuro-fuzzy-fractal approach to the problem of quality control in the manufacturing of sound speakers in a real plant. The quality control of the speakers was done before by manually checking the quality of sound achieved after production (Dickason, 1997). A human expert evaluates the quality of sound of the speakers to decide if production quality was achieved. Of course, this manual checking of the speakers is time consuming and occasionally was the cause of error in quality evaluation (Loctite, 1999). For this reason, it was necessary to consider automating the quality control of the sound speakers. The problem of measuring the quality of the sound speakers is as follows:

1) First, we need to extract the real sound signal of the speaker during the testing period after production.
2) Second, we need to compare the real sound signal to the desired sound signal of the speaker, and measure the difference in some way.
3) Third, we need to decide on the quality of the speaker based on the difference found in step 2. If the difference is small enough then the speaker can be considered of good quality, if not then is bad quality.

The first part of the problem was solved by using a multimedia kit that enable us to extract the sound signal as a file, which basically contains 108000 points over a period of time of 3 seconds (this is the time required for testing). We can say that the sound signal is measured as a time series of data points (Castillo & Melin, 2000), which has the basic characteristics of the speaker. The second part of the problem was solved by using a neuro-fuzzy approach to train a fuzzy model with the data from the good quality speakers (Melin & Castillo, 2000). We used the ANFIS approach (Jang, Sun, & Mizutani, 1997) to obtain a Sugeno fuzzy system (Sugeno, & Kang, 1988) with the time series of the ideal speakers. In the ANFIS approach a neural network (Rumelhart, Hinton, & Williams, 1986) is used to adapt the parameters of the fuzzy system with real data of the problem. With this fuzzy model, the time series of other speakers can be used as checking data to evaluate the total error between the real speaker and the desired one. The third part of the problem was solved by using another set of fuzzy rules (Zadeh, 1975), which basically are fuzzy expert rules to decide on the quality of the speakers based on the total checking error obtained in the previous step. Of course, in this case we needed to define membership functions for the error and quality of the product, and the Mamdani reasoning approach was used. We also use as input variable of the fuzzy system the fractal dimension of the sound signal. The fractal dimension (Mandelbrot, 1987) is a measure of the geometrical complexity of an object (in this case, the time series). We tested our neuro-fuzzy-fractal approach for automated quality control during production with real sound speakers with excellent results. Of course, to measure the efficiency of our intelligent system we compared the results of the neuro-fuzzy-fractal approach to the ones by real human experts. The results clearly show that the neuro-fuzzy-fractal approach was better than the manual method because it reduced the time required for testing and also the accuracy was improved slightly. We think that our neuro-fuzzy-fractal approach for quality control can be used for similar problems, with only some minor changes to the structure of the fuzzy system.

10.2. Basic Concepts of Sound Speakers

In any sound system, ultimate quality depends on the speakers (Dickason, 1997). The best recording, encoded on the most advanced storage device and played by a top-of-the-line deck and amplifier, will sound awful if the system is hooked up to poor speakers. A system's speaker is the component that takes the electronic signal stored on things like CDs, tapes and DVD's and turns it back into actual sound that we can hear.

10.2.1 Sound Basics

To understand how speakers work, the first thing you need to do is understand how sound works. Inside your ear is a very thin piece of skin called the eardrum. When your eardrum vibrates, your brain interprets the vibrations as sound. Rapid changes in air pressure are the most common thing to vibrate your eardrum.

An object produces sound when it vibrates in air (sound can also travel through liquids and solids, but air is the transmission medium when we listen to speakers). When something vibrates, it moves the air particles around it. Those air particles in turn move the air particles around them, carrying the pulse of the vibration through the air as more and more particles are pushed farther from the source of the vibration.

To see how this works, let's look at a simple vibrating object -- a bell. When you ring a bell, the metal vibrates -- flexes in and out -- rapidly. When it flexes out on one side, it pushes out on the surrounding air particles on that side. These air particles then collide with the particles in front of them, which collide with the particles in front of them, and so on. When the bell flexes away, it pulls in on these surrounding air particles, creating a drop in pressure that pulls in on more surrounding air particles, which creates another drop in pressure that pulls in particles that are even farther out, and so on. This decreasing of pressure is called rarefaction.

In this way, a vibrating object sends a wave of pressure fluctuation through the atmosphere. When the fluctuation wave reaches your ear, it vibrates the eardrum back and forth. Our brain interprets this motion as sound. We hear different sounds from different vibrating objects because of variations in:

- sound wave frequency -- A higher wave frequency simply means that the air pressure fluctuates faster. We hear this as a higher pitch. When there are fewer fluctuations in a period of time, the pitch is lower.
- air pressure level -- the wave's amplitude -- determines how loud the sound is. Sound waves with greater amplitudes move our ear drums more, and we register this sensation as a higher volume.

A speaker is a device that is optimized to produce accurate fluctuations in air pressure.

A microphone works something like our ears. It has a diaphragm that is vibrated by sound waves in an area. The signal from a microphone gets encoded on a tape or CD as an electrical signal. When you play this signal back on your stereo, the amplifier sends it to the speaker, which re-interprets it into physical vibrations. In the next section, we'll see how the speaker accomplishes this task.

10.2.2 Making Sound

In the last section we saw that sound travels in waves of air pressure fluctuation, and that we hear sounds differently depending on the frequency and amplitude of these waves. We also learned that microphones translate sound waves into electrical signals, which can be encoded onto CDs, tapes, LPs, etc. Players convert this stored information back into an electric current for use in the stereo system.

A speaker is essentially the final translation machine -- the reverse of the microphone. It takes the electrical signal and translates it back into physical vibrations to create sound waves. When everything is working as it should, the speaker produces nearly the same vibrations that the microphone originally recorded and encoded on a tape, CD, LP, etc. Traditional speakers do this with one or more drivers. A driver produces sound waves by rapidly vibrating a flexible cone, or diaphragm. Figure 10.1 shows a typical speaker driver.

Figure 10.1 A typical speaker driver, with a metal basket, heavy permanent magnet and paper diaphragm.

- The cone, usually made of paper, plastic or metal, is attached on the wide end to the suspension, or surround.
- This rim of flexible material allows the cone to move.
- driver's metal frame, called the basket.
- The narrow end of the cone is connected to the voice coil.
- The coil is attached to the basket by the spider, a ring of flexible material. The spider holds the coil in position, but allows it to move freely back and forth.

Some drivers have a dome instead of a cone. A dome is just a diaphragm that extends out instead of tapering in.

The voice coil is a basic electromagnet. An electromagnet is a coil of wire, usually wrapped around a piece of magnetic metal, such as iron. Running electrical current through the wire creates a magnetic field around the coil, magnetizing the metal it is wrapped around. The field acts just like the magnetic field around a permanent magnet: It has a polar orientation -- a "north" end and a "south" end -- and it is attracted to iron objects. But unlike a permanent magnet, in an electromagnet you can alter the orientation of the poles. If you reverse the flow of the current, the north and south ends of the electromagnet switch.

This is exactly what a stereo signal does -- it constantly reverses the flow of electricity. If you've ever hooked up a stereo system, then you know that there are two outputs wires for each speaker -- typically a black one and a red one. Figure 10.2 shows the wire that runs through the speaker system. Essentially, the amplifier is constantly switching the electrical signal, fluctuating between a positive charge and a negative charge on the red wire. Since electrons always flow in the same direction between positively charged particles and negatively charged particles, the current going through the speaker moves one way and then reverses and flows the other way. This alternating current causes the polar orientation of the electromagnet to reverse itself many times a second.

Figure 10.2 The wire that runs through the speaker system connects to two hook-up jacks on the driver.

So how does this fluctuation make the speaker coil move back and forth? The electromagnet is positioned in a constant magnetic field created by a permanent magnet. These two magnets -- the electromagnet and the permanent magnet -- interact with each other as any two magnets do. The positive end of the electromagnet is attracted to the negative pole of the permanent magnetic field, and the negative pole of the electromagnet is repelled by the permanent magnet's negative pole.

When the electromagnet's polar orientation switches, so does the direction of repulsion and attraction. In this way, the alternating current constantly reverses the magnet forces between the voice coil and the permanent magnet. This pushes the coil back and forth rapidly, like a piston. When the electrical current flowing through the voice coil changes direction, the coil's polar orientation reverses. This changes the magnetic forces between the voice coil and the permanent magnet, moving the coil and attached diaphragm back and forth.

When the coil moves, it pushes and pulls on the speaker cone. This vibrates the air in front of the speaker, creating sound waves. The electrical audio signal can also be interpreted as a wave. The frequency and amplitude of this wave, which represent the original sound wave, dictates the rate and distance that the voice coil moves. This, in turn, determines the frequency and amplitude of the sound waves produced by the diaphragm. Different driver sizes are better suited for certain frequency ranges. For this reason, loudspeaker units typically divide a wide frequency range among multiple drivers. In the next section, we'll find out how speakers divide up the frequency range, and we'll look at the main driver types used in loudspeakers.

10.2.3 Chunks of the Frequency Range

In the last section we saw that traditional speakers produce sound by pushing and pulling an electromagnet attached to a flexible cone. Although drivers all work on the same concept, there is actually a wide variety in driver size and power. The basic driver types are: 1) Woofers 2) Tweeters 3) Midrange (Figure 10.3).

Woofers are the biggest drivers, and are designed to produce low frequency sounds. Tweeters are much smaller units, designed to produce the highest frequencies. Midrange speakers produce a range of frequencies in the middle of the sound spectrum.

And if you think about it, this makes perfect sense. To create higher frequency waves -- waves in which the points of high pressure and low pressure are closer together -- the driver diaphragm must vibrate more quickly. This is harder to do with a large cone because of the mass of the cone. Conversely, it's harder to get a small driver to vibrate slowly enough to produce very low frequency sounds. It's more suited to rapid movement.

To produce quality sound over a wide frequency range more effectively, you can break the entire range into smaller chunks that are handled by specialized drivers. Quality loudspeakers will typically have a woofer, a tweeter and sometimes a midrange driver, all included in one enclosure.

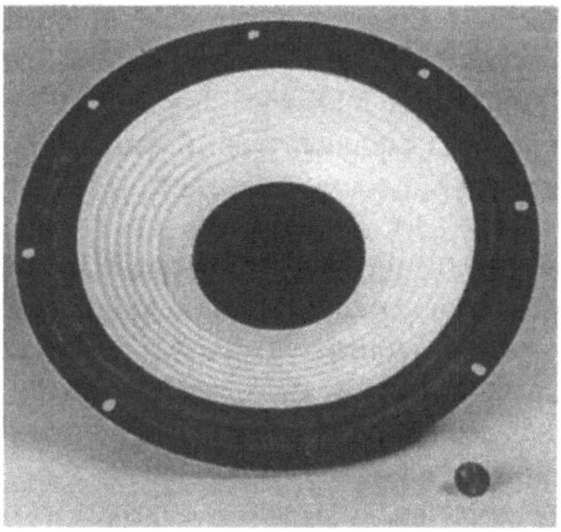

Woofer

Tweeter

Midrange

Figure 10.3 Three basic driver types: woofer, tweeter, and midrange.

Of course, to dedicate each driver to a particular frequency range, the speaker system first needs to break the audio signal into different pieces -- low frequency, high frequency and sometimes mid-range frequencies. This is the job of the speaker crossover. The most common type of crossover is passive, meaning it doesn't need an external power source because it is activated by the audio signal passing through it. This sort of crossover uses inductors, capacitors, and sometimes, other circuitry components. Capacitors and inductors only become good conductors under certain conditions. A crossover capacitor will conduct the current very well when the frequency exceeds a certain level, but will conduct poorly when the frequency is below that level. A crossover inductor acts in the reverse manner -- it is only a good conductor when the frequency is below a certain level (Figure 10.4).

When the electrical audio signal travels through the speaker wire to the speaker, it passes through the crossover units for each driver. To flow to the tweeter, the current will have to pass through a capacitor. So for the most part, the high frequency part of the signal will flow on to the tweeter voice coil. To flow to the woofer, the current passes through an inductor, so the driver will mainly respond to low frequencies. A crossover for the mid-range driver will conduct the current through a capacitor and an inductor, to set an upper and lower cutoff point.

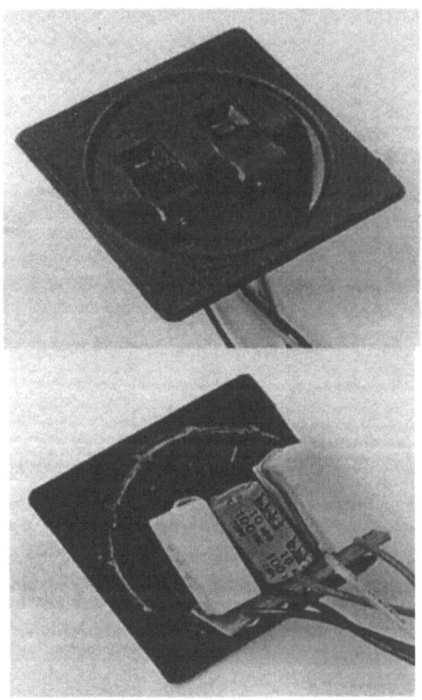

Figure 10.4 Typical crossover unit from a loudspeaker.

There are also active crossovers. Active crossovers are electronic devices that pick out the different frequency ranges in an audio signal before it goes on to the amplifier (you use an amplifier circuit for each driver). They have several advantages over passive crossovers, the main one being that you can easily adjust the frequency ranges. Passive crossover ranges are determined by the individual circuitry components -- to change them, you need to install new capacitors and inductors. Active crossovers aren't as widely used as passive crossovers, however, because the equipment is much more expensive and you need multiple amplifier outputs for your speakers.

Crossovers and drivers can be installed as separate components in a sound system, but most people end up buying speaker units that house the crossover and multiple drivers in one box. In the next section, we'll find out what these speaker enclosures do and how they affect the speaker's sound quality.

10.2.4 Boxes of Sound

In most loudspeaker systems, the drivers and the crossover are housed in some sort of speaker enclosure. These enclosures serve a number of functions. On their most basic level, they make it much easier to set up the speakers. Everything's in one unit and the drivers are kept in the right position, so they work together to produce the best sound. They are usually built with heavy wood or another solid material that will effectively absorb the driver's vibration. If you simply placed a driver on a table, the table would vibrate so much it would drown out a lot of the speaker's sound.

Additionally, the speaker enclosure affects how sound is produced. When we looked at speaker drivers, we focused on how the vibrating diaphragm emitted sound waves in front of the cone. But, since the diaphragm is moving back and forth, it's actually producing sound waves behind the cone as well. Different enclosure types have different ways of handling these "backward" waves (Figure 10.5).

The most common type of enclosure is the sealed enclosure, also called acoustic suspension enclosure. These enclosures are completely sealed, so no air can escape. This means the forward wave travels outward into the room, while the backward wave travels only into the box. Of course, since no air can escape, the internal air pressure is constantly changing -- when the driver moves in, the pressure is increased, and when the driver moves out, it is decreased. Both movements create pressure differences between the air inside the box and the air outside the box. The air will always move to equalize pressure levels, so the driver is constantly being pushed toward its "resting" state -- the position at which internal and external air pressure are the same (Figure 10.6).

Figure 10. 5 A typical sealed speaker enclosure that holds a tweeter, a woofer and a midrange driver.

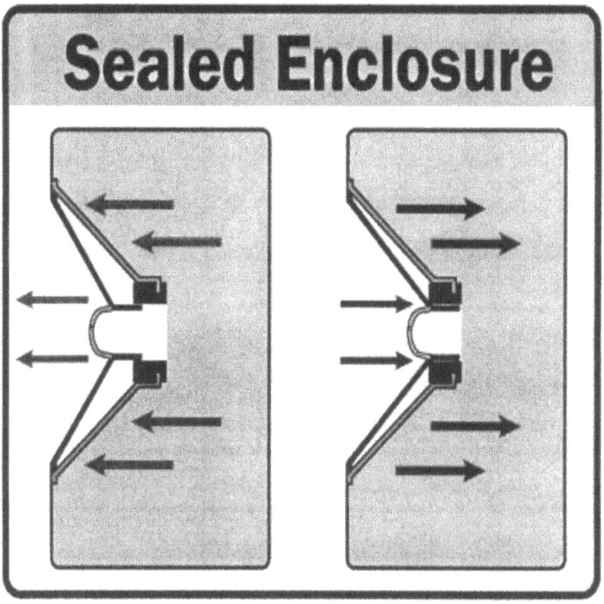

Figure 10.6 In a sealed speaker setup, the driver diaphragm compresses air in the enclosure when it moves in, and rarefies air when it moves out.

These enclosures are less efficient than other designs because the amplifier has to boost the electrical signal to overcome the force of air pressure. The force serves a valuable function, however -- it acts like a spring to keep the driver in the right position. This makes for tighter, more precise sound production.

The other thing you can do with the backward wave is to redirect it outward, using it to supplement the forward sound wave. The most common way to do this is to build a small port into the speaker. In these bass reflex speakers, the backward motion of the diaphragm pushes sound waves out of the port, boosting the overall sound level (Figure 10.7). The main advantage of bass reflex enclosures is efficiency. The power moving the driver is used to emit two sound waves rather than one. The disadvantage is that there is no air pressure difference to spring the driver back into place, so the sound production is not as precise.

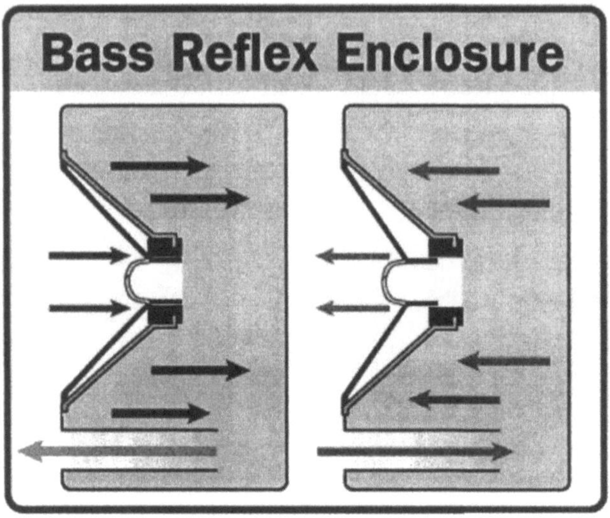

Figure 10.7 A bass reflex speaker produces two sound waves by moving one driver. When the driver compresses air forward, it rarefies it backward, and vice versa. The second sound wave is emitted from a port at the base of the speaker enclosure.

Passive radiator enclosures are very similar to bass reflex units, but in passive radiator enclosures, the backward wave moves an additional, passive driver, instead of escaping out of the port. The passive driver is just like the main, active drivers except it doesn't have an electromagnet voice coil, and it isn't connected to the amplifier. It is moved only by the sound waves coming from the active drivers. This type of enclosure is more efficient than sealed designs and more precise than bass reflex models (Figure 10.8).

Some enclosure designs have an active driver facing one way and a passive driver facing the other way. This dipole design diffuses the sound in all directions, making it a good choice for the rear channels in a home theater system.

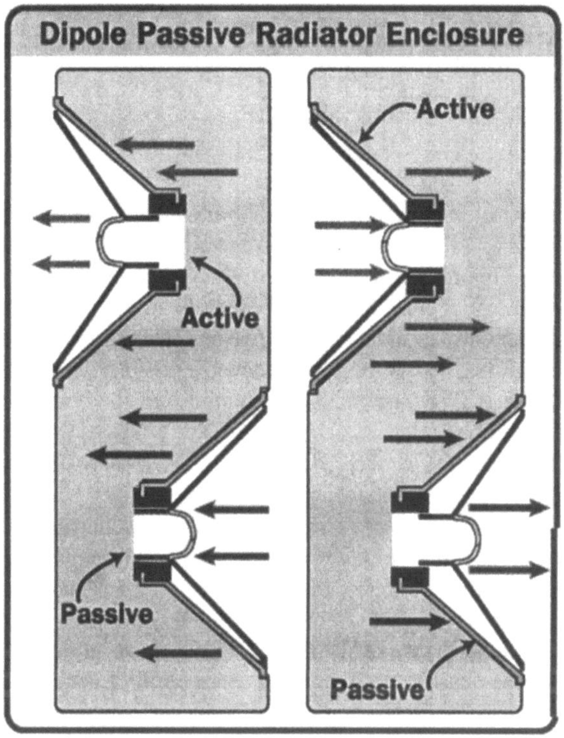

Figure 10.8 The backward air compression and rarefaction caused by the active driver push and pull on the passive driver. A speaker with a dipole design emits sound waves in both directions.

These are just a few of the many enclosure types available. There are a huge range of speaker units on the market, with a variety of unique structures and driver arrangements.

10.2.5 Alternative Speaker Designs

Most loudspeakers produce sound with traditional drivers. But there are a few other technologies on the market. These designs have some advantages over traditional dynamic speakers, but they fall short in other areas. For this reason, they are often used in conjunction with driver units.

The most popular alternative is the electrostatic speaker. These speakers vibrate air with a large, thin, conductive diaphragm panel. This diaphragm panel is suspended between two stationary conductive panels that are charged with electrical current from a wall outlet. These panels create an electrical field with a positive end and a negative end. The audio signal runs a current through the suspended panel, rapidly switching between a positive charge and a negative charge. When the charge is positive, the panel is drawn toward the negative end of the field, and when the charge is negative, it moves toward the positive end in the field.

In this way, the diaphragm rapidly vibrates the air in front of it. Because the panel has such a low mass, it responds very quickly and precisely to changes in the audio signal. This makes for clear, extremely accurate sound reproduction. The panel doesn't move a great distance, however, so it is not very effective at producing lower frequency sounds. For this reason, electrostatic speakers are often paired with a woofer that boosts the low frequency range. The other problem with electrostatic speakers is that they must be plugged into the wall and so are more difficult to place in a room.

Another alternative is the planar magnetic speaker. These units use a long, metal ribbon suspended between two magnetic panels. They basically work the same way as electrostatic speakers, except that the alternating positive and negative current moves the diaphragm in a magnetic field rather than an electric field. Like electrostatic speakers, they produce high-frequency sound with extraordinary precision, but low frequency sounds are less defined. For this reason, the planar magnetic speaker is usually used only as a tweeter.

Both of these designs are becoming more popular with audio enthusiasts. But traditional dynamic drivers are still the most prevalent technology, far and away. You'll find them everywhere you go -- not only in stereo setups, but also in alarm clocks, public address systems, televisions, computers, headphones and tons of other devices.

10.3 Description of the Problem

The basic problem consists in the identification of sound signal quality. Of course, this requires a comparison between the real measured sound signal and the ideal good sound signal. We need to be able to accept speakers, which have a sound signal that do not differ much from the ideal signals. We show in Figure 10.9 the form of the sound signal for a good speaker (of a specific type). The measured signal contains about 108 000 points in about 3 seconds. We need to compare any other measured signal with the good one and calculate the total difference between both of them, and if the difference is small then we can accept the speaker as a good one. On the other hand, if the difference is large then we reject the speaker as a bad one.

We show in Figure 10.10 the sound signal for a speaker of bad quality. We can clearly see the difference in the geometrical form of this signal and the one shown in Figure 10.9. In this case, the difference between the figures is sufficiently large and we easily determine that the speaker is of bad quality. We also show in Figure 10.11 another sound signal for a bad quality speaker. Again, we can see clearly the difference in the form of the signal with respect to the good speaker.

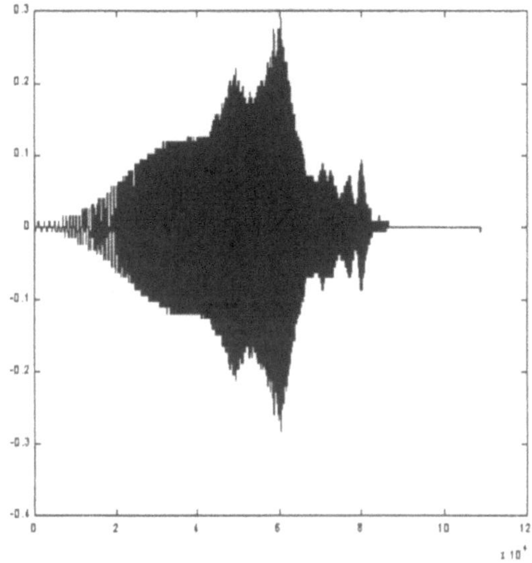

Figure 10.9 Sound signal of a Good Speaker.

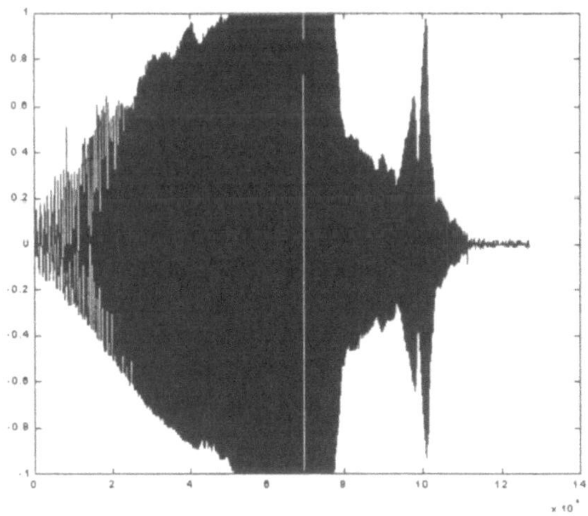

Figure 10.10 Sound signal of Bad Speaker (Case 1)

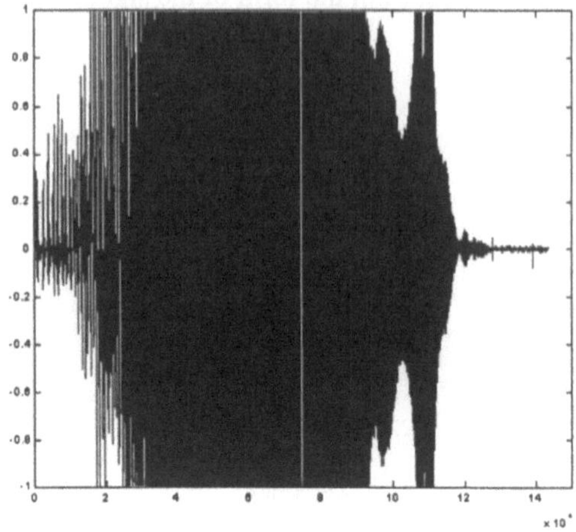

Figure 10.11 Sound signal of Bad Speaker (Case 2).

10.4 Fractal Dimension of a Sound Signal

Recently, considerable progress has been made in understanding the complexity of an object through the application of fractal concepts (Mandelbrot, 1987) and dynamic scaling theory (Rasband, 1990). For example, financial time series show scaled properties suggesting a fractal structure (Castillo & Melin, 1998, 1999). The fractal dimension of a geometrical object can be defined as follows:

$$d = \lim_{r \to 0} [\ln N(r)] / [\ln(1/r)] \tag{10.1}$$

where N(r) is the number of boxes covering the object and r is the size of the box. An approximation to the fractal dimension can be obtained by counting the number of boxes covering the boundary of the object for different r sizes and then performing a logarithmic regression to obtain d (box counting algorithm). In Figure 10.12, we illustrate the box counting algorithm for a hypothetical curve C. Counting the number of boxes for different sizes of r and performing a logarithmic linear regression, we can estimate the box dimension of a geometrical object with the following equation:

$$\ln N(r) = \ln\beta - d \ln r \tag{10.2}$$

this algorithm is illustrated in Figure 10.13.

The fractal dimension can be used to characterize an arbitrary object. The reason for this is that the fractal dimension measures the geometrical complexity of objects. In this case, a time series can be classified by using the numeric value of the fractal dimension (d is between 1 and 2 because we are on the plane xy). The reasoning behind this classification scheme is that when the boundary is smooth the fractal dimension of the object will be close to one. On the other hand, when the boundary is rougher the fractal dimension will be close to a value of two.

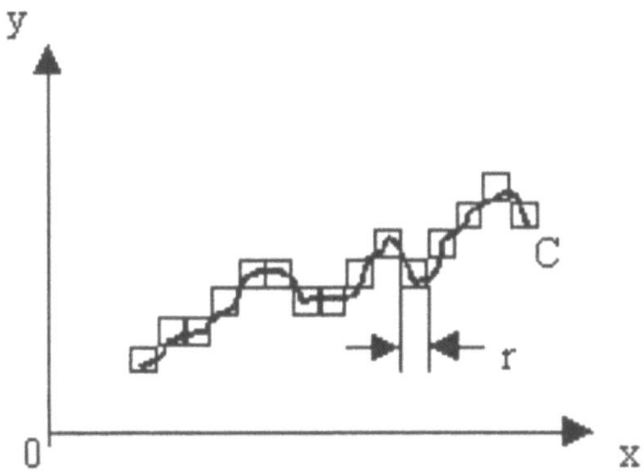

Figure 10.12 Box counting algorithm for a curve C.

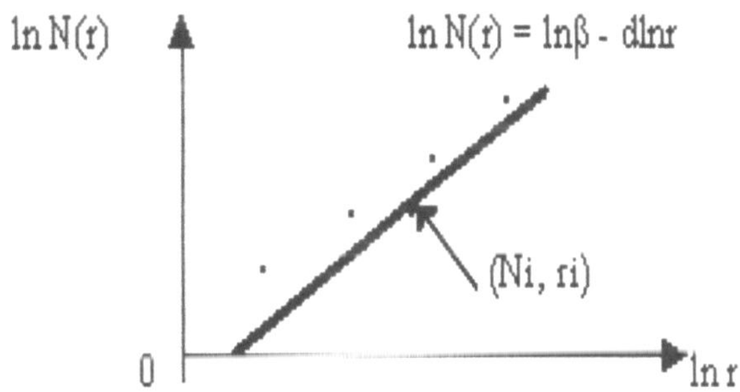

Figure 10.13 Logarithmic regression to find dimension.

We developed a computer program in MATLAB for calculating the fractal dimension of a sound signal. The computer program uses as input the figure of the signal and counts the number of boxes covering the object for different grid sizes. The fractal dimension for the sound signal of Figure 10.9 is of 1.6479, which is a low value because it corresponds to a good speaker. On the other hand, the fractal dimension for Figure 10.10 is 1.7843, which is a high value (bad speaker). Also, for the case of Figure 10.11 the dimension is 1.8030, which is even higher (again, a bad speaker).

We show in Figure 10.14 an illustration of the box counting algorithm for the sound signal of a good speaker. In this figure we can appreciate how we can count the number of boxes by using the sum of the 1's in the grid.

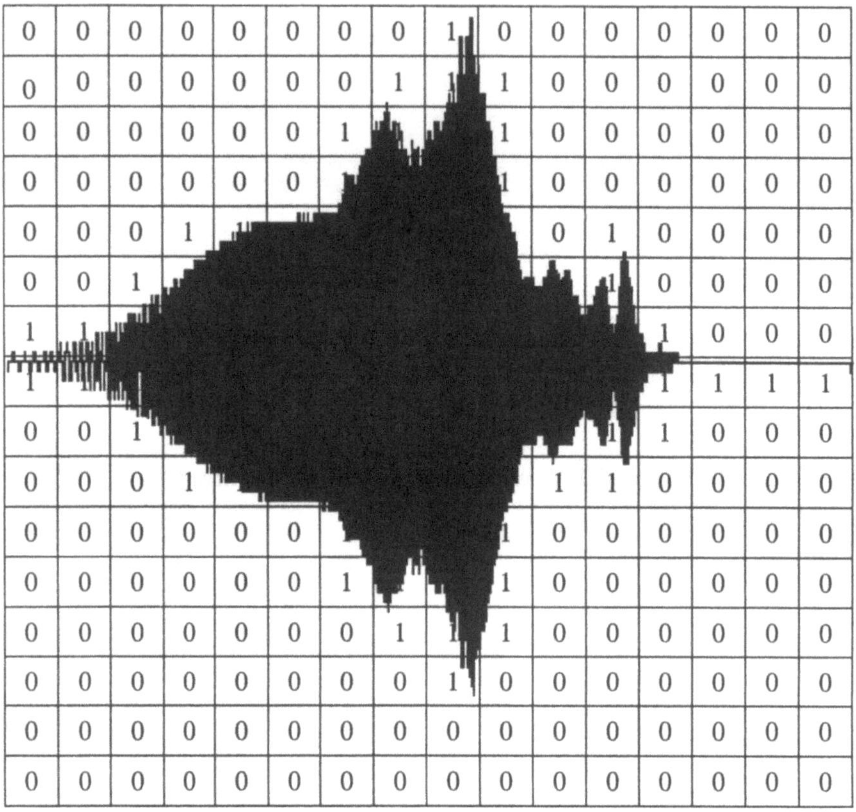

Figure 10.14 A sample grid to count the number of boxes covering a sound signal.

10.5. Experimental Results

We describe in this section the experimental results obtained with the intelligent system for automated quality control. The intelligent system uses a fuzzy rule base to determine automatically the quality of sound in speakers. We used a neuro-fuzzy approach to adapt the parameters of the fuzzy system using real data from the problem. We used the time series of 108000 points measured from a good sound speaker (in a period of 3 seconds) as training data in the ANFIS approach. We then use the measured data of any other speaker as checking data, to compare the form of the sound signals. We show in Figures 10.15 and 10.16 two cases where the ANFIS approach is used to adapt a fuzzy system with training data of good sound speakers. The approximation is very good considering the complexity of the problem. Once the training was done, we used the fuzzy system for measuring the total difference between a given signal and the good ones. This difference is used to decide on the final quality of the speaker using another set of fuzzy rules with the Mamdani approach. The fuzzy rules are as follows:

IF Difference is small AND Fractal Dimension is small
THEN Quality is Excellent
IF Difference is regular AND Fractal Dimension is small
THEN Quality is Good
IF Difference is regular AND Fractal Dimension is high
THEN Quality is Medium
IF Difference is medium AND Fractal Dimension is small
THEN Quality is Medium
IF Difference is medium AND Fractal Dimension is high
THEN Quality is Bad
IF Difference is large AND Fractal Dimension is small
THEN Quality is Medium
IF Difference is large AND Fractal Dimension is high
THEN Quality is Bad
IF Difference is small AND Fractal Dimension is high
THEN Quality is Medium

We show in Figure 10.17 the implementation of the fuzzy rule base, in the Fuzzy Logic Toolbox of the MATLAB programming language. We also show in Figure 10.18 the non-linear surface for modeling the quality control problem. Finally, we show in Figure 10.19 the fuzzy rule viewer of MATLAB, in which the fuzzy rule base, is used for specific values of the input variables.

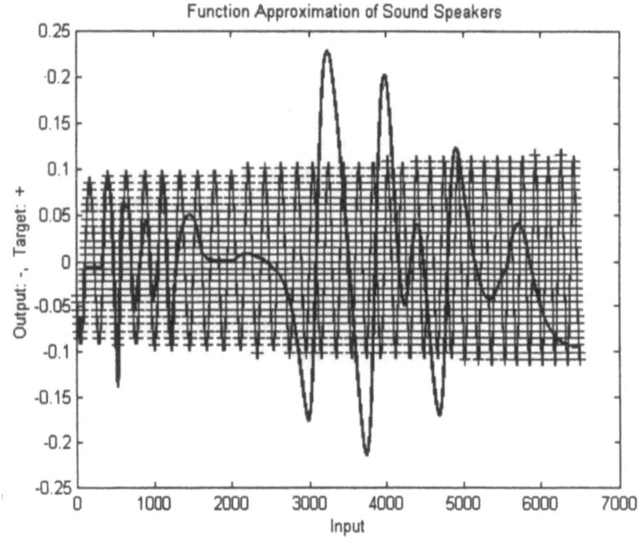

Figure 10.15 Function approximation of the sound signal using ANFIS (Case 1).

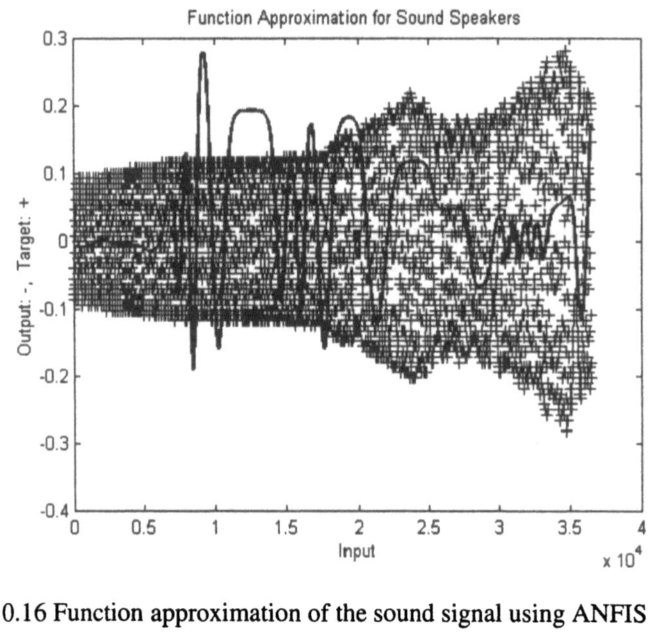

Figure 10.16 Function approximation of the sound signal using ANFIS (Case 2).

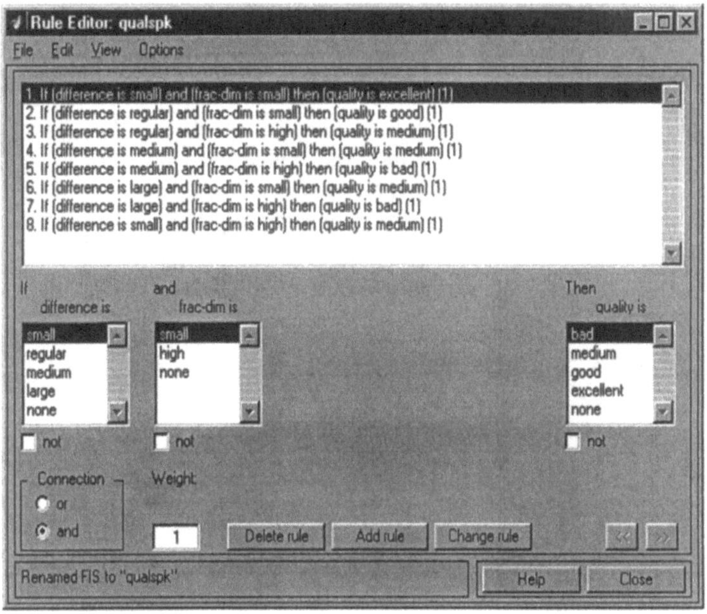

Figure 10.17 Fuzzy rule base for quality control.

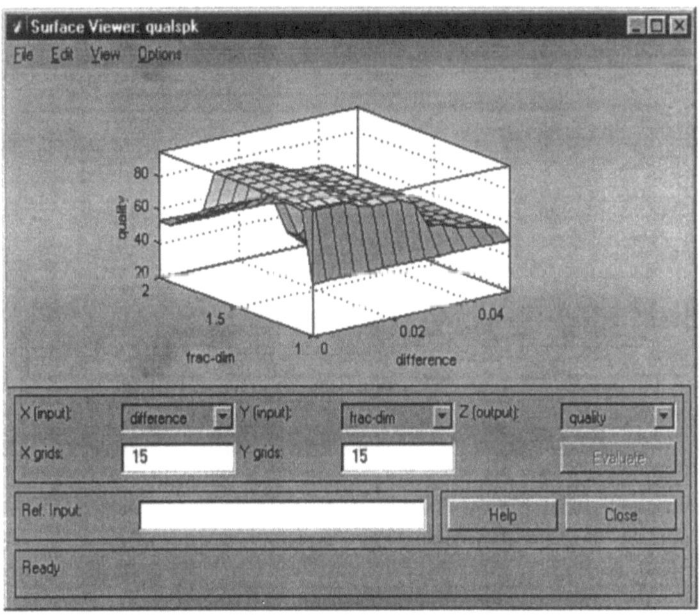

Figure 10.18 Non-linear surface for quality control.

In the fuzzy rule base, the "Difference", "Fractal Dimension" and "Quality" are considered as linguistic variables in the fuzzy rules. We used gaussian membership functions for the linguistic values of all the variables. The membership functions for the linguistic values of these variables were tuned by using a simple genetic algorithm. Of course, the human experts on quality evaluation gave initial values for the fuzzy system, but these were optimized using real data for the problem and by applying the simple genetic algorithm.

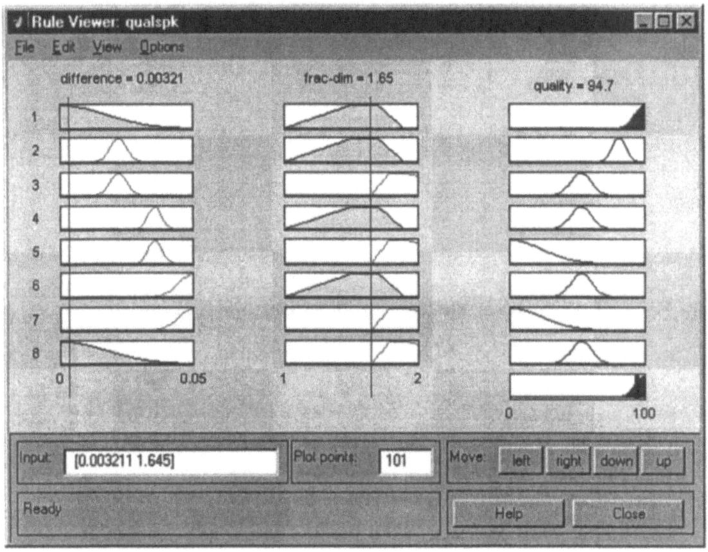

Figure 10.19 Rule viewer for specific input values.

10.6 Summary

We described in this chapter the application of a neuro-fuzzy-fractal approach to the problem of automating the quality control of sound speakers during manufacturing in a real plant. We have implemented an intelligent system for quality control in the MATLAB programming language using the ANFIS approach. We also use the fractal dimension as a measure of geometrical complexity of the sound signals. The intelligent system performs rather well considering the complexity of the problem. The intelligent system has been tested in a real manufacturing plant with very good results. We think that our approach for automating quality evaluation can be used for similar problems with only minor changes in the membership functions and rules.

Chapter 11

Intelligent Manufacturing of Television Sets

We describe in this chapter an intelligent system for controlling the electrical tuning process during the manufacturing of television sets. The electrical tuning problem consists in controlling the imaging system of the television set to meet production quality requirements. Traditionally, this tuning process has been performed by human operators by manually adjusting the imaging system in the television set. In our approach, we use fuzzy logic to automate the tuning process. We use a fuzzy system for controlling the voltage, current, and the time during the tuning process, so that the best possible quality of image is achieved. We also use a specific genetic algorithm for optimizing the parameters and number of rules of the fuzzy system. We have implemented this intelligent system for control in the MATLAB programming language with good simulation results. We compared the fuzzy-genetic approach with other methods for control with very good results. The intelligent system was also tested in a real production plant for several months with very good results.

11.1 Introduction

We describe in this chapter the application of fuzzy logic techniques to the problem of automating the electrical tuning process during the manufacturing of television sets. The electrical tuning process consists in adjusting the voltage, current intensity, and time to produce the best image possible. We have designed a set of fuzzy if-then rules that contains the knowledge of the human experts in the electrical tuning process of television sets. With these fuzzy rules we have developed an intelligent system for automated electrical tuning of television sets during the manufacturing process. We used a genetic approach to adapt (learn) the parameter values of the fuzzy system according to real data about the problem

The basic problem that we are considering in this chapter is how to artificially reproduce images in the best way possible in devices like television sets or monitors. While producing television sets or monitors in a plant, we normally have a section in charge of adjusting the imaging system of these devices. Traditionally, a human expert operator adjusts the imaging system by experience using a special remote control and the measurements of voltage and current intensity. We are now considering an intelligent system for controlling the electrical tuning process of television sets during production. The intelligent system has a knowledge base, consisting of fuzzy if then rules, that contains the expert knowledge about tuning the imaging system of television sets. The main reason for using fuzzy logic is that we need to represent and also reason with uncertainty in this application (Castillo & Melin, 2001). We need to use voltage, current intensity, time, and quality as linguistic variables in the fuzzy rules, and define the membership functions for these variables according to real data about the problem and the knowledge of the experts. We begin this chapter by first describing the basic concepts of the imaging system of a television set. Then, we describe how to use fuzzy logic techniques to automate the electrical tuning process of television sets and show some simulation results. We also show some details about the genetic algorithm that was used for optimizing the fuzzy system (Jang, Sun, & Mizutani, 1997). Finally, we give some concluding remarks and mention future work in this area.

11.2 Imaging System of the Television Sets

We describe in this section the basic concepts to understand the imaging system in a television. We begin by describing the cathode ray tube, then we describe the principles of the black and white television set, and then the basic principles of the color television set. We consider this material necessary to appreciate the use of soft computing techniques in solving the problem of controlling the imaging system of the television set.

11.2.1 The Cathode Ray Tube

Almost all television sets in use today rely on a device known as the cathode ray tube, or CRT, to display their images. The LCDs and plasma displays are sometimes seen, but they are still rare when compared to CRTs. It is even possible to make a television screen out of thousands of ordinary 60-watt light bulbs! You may have seen something like this at an outdoor event like a football game. Let's start with the CRT, however, because CRTs are the most common way of displaying images today.

The terms anode and cathode are used in electronics as synonyms for positive and negative terminals. For example, you could refer to the positive terminal of a battery as the anode and the negative terminal as the cathode. In a cathode ray tube, the "cathode" is a heated filament (not unlike the filament in a normal light bulb). The heated filament exists in a vacuum created inside a glass "tube." The "ray" is a stream of electrons that naturally pour off a heated cathode into the vacuum. Electrons are negative. The anode is positive, so it attracts the electrons pouring off of the cathode. In a cathode ray tube of a television set, the stream of electrons is focused by a focusing anode into a tight beam and then accelerated by an accelerating anode. This tight, high-speed beam of electrons flies through the vacuum in the tube and hits the flat screen at the other end of the tube. This screen is coated with phosphor, which glows when struck by the beam. Figure 11.1 shows this situation.

As you can see in this figure, there's not a whole lot to a basic cathode ray tube. There is a cathode and a pair (or more) of anodes. There is the phosphor coated screen. There is a conductive coating inside the tube to soak up the electrons that pile up at the screen-end of the tube. However, you can see in this diagram no way to "steer" the beam -- the beam will always land in a tiny dot right in the center of the screen. That's why, if you look inside any television set, you will find that the tube is wrapped in coils of wires. The following picture gives you a view of a typical set of steering coils: Figure 11.2 shows this picture. The steering coils are simply copper windings. These coils are able to create magnetic fields inside the tube, and the electron beam responds to the fields. One set of coils creates a magnetic field that moves the electron beam vertically, while another set moves the beam horizontally. By controlling the voltages in the coils, you can position the electron beam at any point on the screen.

11.2.2 Phosphor

A phosphor is any material that, when exposed to radiation, emits visible light. The radiation might be ultraviolet light or a beam of electrons. Any fluorescent color is really a phosphor - fluorescent colors absorb invisible (to us) ultraviolet light and emit visible light at a characteristic color.

In a CRT, phosphor coats the inside of the screen. When the electron beam strikes the phosphor, it makes the screen glow. In a black-and-white screen there is one phosphor that glows white when struck. In a color screen there are three phosphors arranged as dots or stripes that emit red, green and blue light. There are also three electron beams to illuminate the three different colors together.

There are thousands of different phosphors that have been formulated. They are characterized by their emission color and the length of time emission lasts after they are excited.

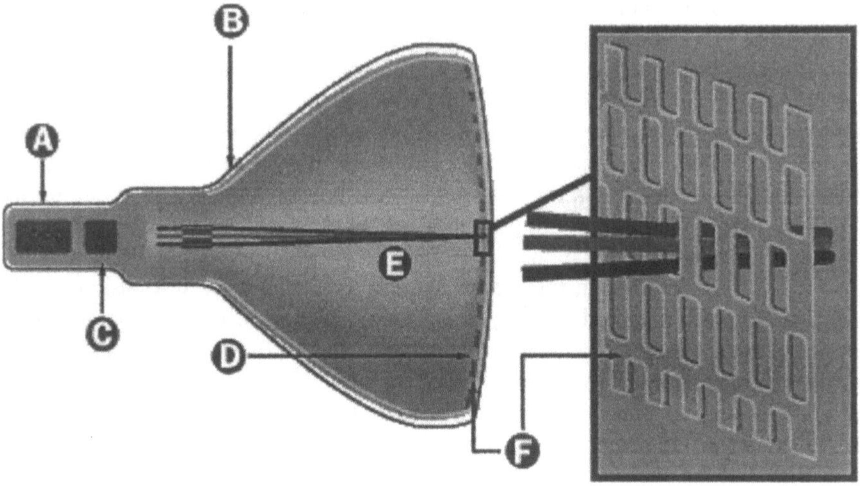

A Cathode B Conductive coating C Anode D Phosphor-coated screen E Electron beams F Shadow mask

Figure 11.1 Basic architecture of a color television set.

Figure 11.2 Typical set of steering coils inside a television set.

11.2.3 The Black-and-White Television Signal

In a black-and-white television set, the screen is coated with white phosphor and the electron beam "paints" an image onto the screen by moving the electron beam across the phosphor a line at a time. To "paint" the entire screen, electronic circuits inside the television set use the magnetic coils to move the electron beam in a "raster scan" pattern across and down the screen. The beam paints one line across the screen from left to right. It then quickly flies back to the left side, moves down slightly and paints another horizontal line, and so on down the screen. Figure 11.3 shows the black and white television set.

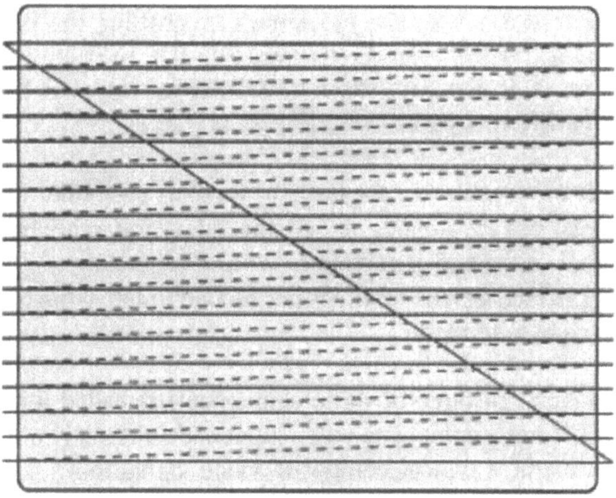

Figure 11.3 Black and white television set.

In this figure 11.3, the blue lines represent lines that the electron beam is "painting" on the screen from left to right, while the red dashed lines represent the beam flying back to the left. When the beam reaches the right side of the bottom line, it has to move back to the upper left corner of the screen, as represented by the green line in the figure. When the beam is "painting" it is on, and when it is flying back it is off so that it does not leave a trail on the screen. The term horizontal retrace is used to refer to the beam moving back to the left at the end of each line, while the term vertical retrace refers to its movement from bottom to top. As the beam paints each line from left to right, the intensity of the beam is changed to create different shades of black, gray and white across the screen. Because the lines are spaced very closely together, your brain integrates them into a single image. A television set screen normally has 525 lines visible from top to bottom.

All television sets use an interlacing technique when painting the screen. In this technique, the screen is painted 60 times per second but only half the lines are painted per frame. The beam paints every other line as it moves down the screen, for example every odd-numbered line. Then the next time it moves down the screen it paints the even-numbered lines, alternating back and forth between even-numbered and odd-numbered lines on each pass. The entire screen, in two passes, is painted 30 times every second. The alternative to interlacing is called progressive scanning, which paints every line on the screen 60 times a second. Most computer monitors use progressive scanning because it significantly reduces flicker.

Because the electron beam is painting all 525 lines 30 times per second, it paints a total of 15,750 lines per second. (Some people can actually hear this frequency as a very high-pitched sound emitted when the television set is on.)

When a television station wants to broadcast a signal to your television set, or when your VCR wants to display the movie on a video tape on your television set, the signal needs to mesh with the electronics controlling the beam so that the television set can accurately paint the picture that the TV station or VCR sends. The TV station or VCR therefore sends a well-known signal to the television set that contains three different parts:

- Intensity information for the beam as it paints each line
- Horizontal retrace signals to tell the television set when to move the beam back at the end of each line
- Vertical retrace signals 60 times per second to move the beam from bottom right to top left.

A signal that contains all three of these components is called a composite video signal. If your VCR has a composite video input capability, it is normally a yellow RCA jack. One line of a typical composite video signal looks something like in Figure 11.4.

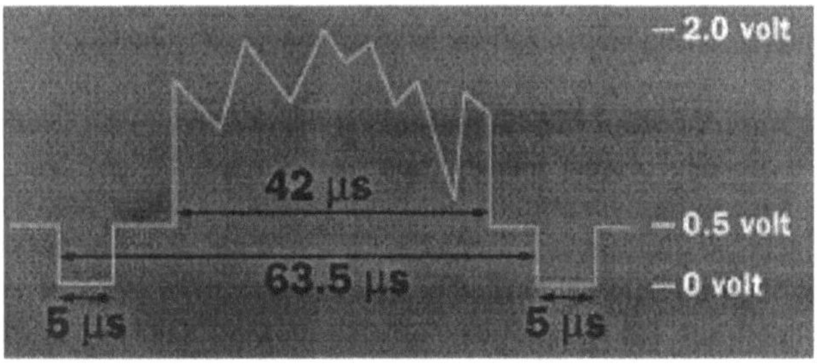

Figure 11.4. Composite video signal.

The horizontal retrace signals are five microsecond (abbreviated as "us" in the figure) pulses at zero volts. Electronics inside the television set can detect these pulses and trigger the beam's horizontal retrace with them. The actual signal for the line is a varying wave between 0.5 volts and 2.0 volts, with 0.5 volts representing black and 2 volts representing white. This signal drives the intensity circuit for the electron beam. In a black-and-white television set this signal can consume about 3.5 MHz of bandwidth, while in a color set the limit is about 3.0 MHz.

A vertical retrace pulse is similar to a horizontal retrace pulse but is 400 to 500 microseconds long. The vertical retrace pulse is serrated with horizontal retrace pulses in order to keep the horizontal retrace circuit in the television set synchronized.

11.2.4 Adding Color

A color television set screen differs from a black-and-white screen in three ways:

1. There are three electron beams that move simultaneously across the screen. They are named the red, green and blue beams.
2. The screen is not coated with a single sheet of phosphor as in a black and white television set. Instead the screen is coated with red, green and blue phosphors arranged in dots or stripes. If you turn on your television set or computer monitor and look closely at the screen with a magnifying glass, you will be able to see the dots.
3. On the inside of the tube, very close to the phosphor coating, there is a thin metal screen called a shadow mask. This mask is perforated with very small holes that are aligned with the phosphor dots on the screen.

The following diagram (Figure 11.5) shows how the shadow mask works.

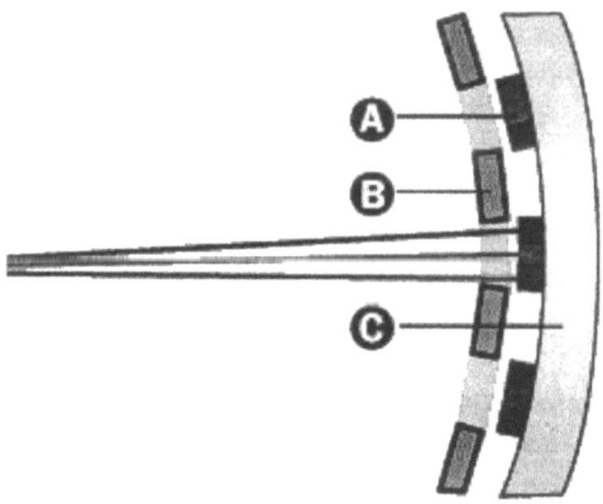

A Phosphors B Shadow mask C Glass

Figure 11.5 Diagram showing how the shadow mask works.

When a color television set needs to create a red dot, it fires the red beam at the red phosphor. Similarly for green and blue dots. To create a white dot, red, green and blue beams are fired simultaneously -- the three colors mix together to create white. To create a black dot, all three beams are turned off as they scan past the dot. All other colors on a television screen are combinations of red, green and blue.

A color television signal starts off looking just like a black-and-white signal. An extra chrominance signal is added by superimposing a 3.579545 MHz sine wave onto the standard black-and-white signal. Right after the horizontal sync pulse, eight cycles of a 3.579545 MHz sine wave are added as a color burst. Figure 11.6

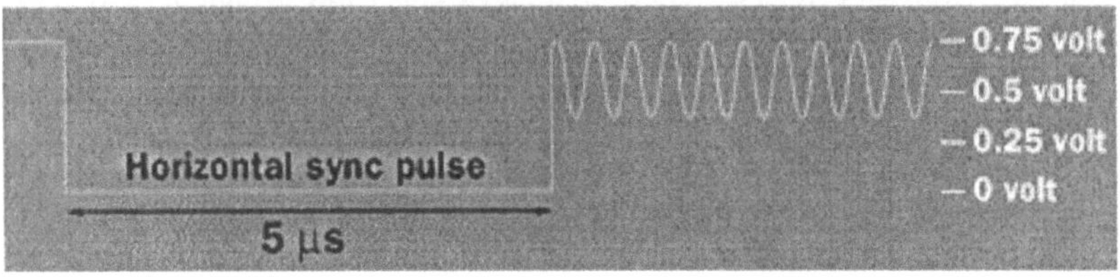

Figure 11.6 Color television signal.

Following these eight cycles, a phase shift in the chrominance signal indicates the color to display. The amplitude of the signal determines the saturation. The following table 11.1 shows you the relationship between color and phase:

Table 11.1. Relationship between color and phase.

Color	*Phase*
Burst	0 degrees
Yellow	15 degrees
Red	75 degrees
Magenta	135 degrees
Blue	195 degrees
Cyan	255 degrees
Green	315 degrees

A black-and-white television set filters out and ignores the chrominance signal. A color television set picks it out of the signal and decodes it, along with the normal intensity signal, to determine how to modulate the three color beams.

Finally, we end this section with some remarks about the human eye and the way different colors are appreciated. First, in Figure 11.7 we show the different percentages of color that a human needs to appreciate "white". We need a combination of 30% red, 59% green, and 11% of blue to see the white color. Similarly, any other color results from a combination of red, green, and blue. We also show in Figure 11.8 the characteristic wavelength and intensity of light for the human eye.

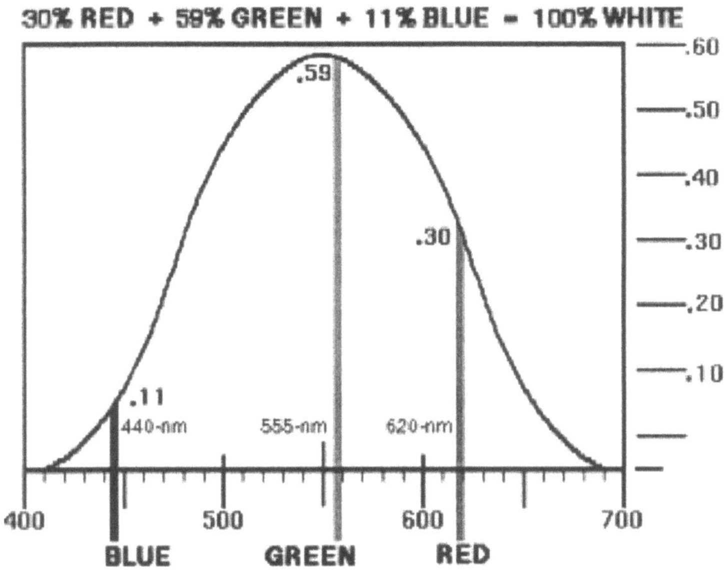

Figure 11.7 Combination of red, green and blue to achieve "white" color.

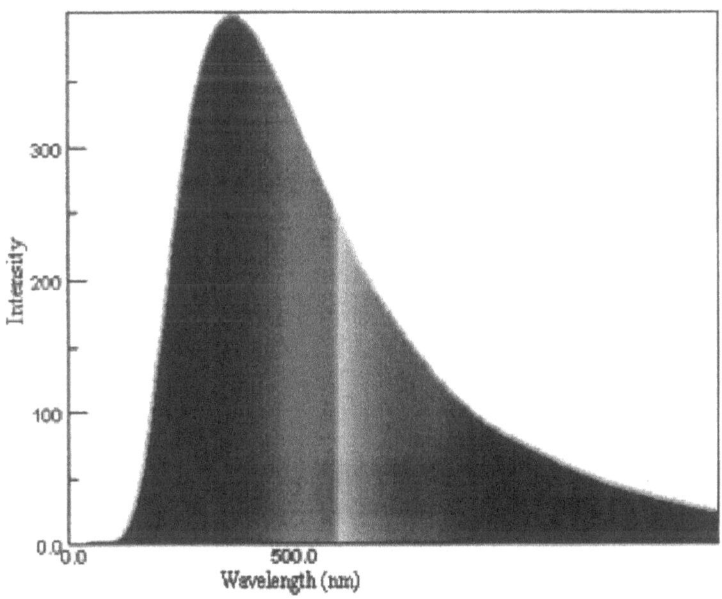

Figure 11.8 Wavelength and intensity of light for the human eye.

11.3 Breeder Genetic Algorithm for Optimization

In this chapter, we used a floating point genetic algorithm (Kuri, 1999) to adjust the parameter vector θ, specifically we used the Breeder Genetic Algorithm (BGA). The genetic algorithm is used to optimize the fuzzy system for control that will be described later. A BGA can be described by the following equation

$$BGA=(P_g^0, N, T, \Gamma, \Delta, HC, F, term) \quad (11.1)$$

where: P_g^0=initial population, N=the size of the population, T=the truncation threshold, Γ=the recombination operator, Δ=the mutation operator, HC=the hill climbing method, F=the fitness function, *term*=the termination criterion.

The BGA uses a selection scheme called truncation selection. The %T best individuals are selected and mated randomly until the number of offspring is equal the size of the population. The offspring generation is equal to the size of the population. The offspring generation replaces the parent population. The best individual found so far will remain in the population. Self-mating is prohibited (Mühlenbein, & Schilierkamp, 1993). As a recombination operator we used "extended intermediate recombination", defined as: If $x=(x_i,x_n)$ *and* y $y=(y_1,...,y_n)$ are the parents, then the successor $z=(z_1,...,z_n)$ is calculated by:

$$z_i=x_i+\alpha_i(y_i-x_i) \qquad i=1,...n \quad (11.2)$$

The mutation operator is defined as follows: A variable x_i is selected with probability p_m for mutation. The BGA normally uses $p_m = 1/n$. At least one variable will be mutated. A value out of the interval $[-range_i, range_i]$ is added to the variable. $range_i$ defines the mutation range. It is normally set to $(0.1 \times searchinterval_i)$. searchinterval$_i$ is the domain of definition for variable x_i. The new value z_i is computed according to

$$z_i=x_i \pm range_i \cdot \delta \quad (11.3)$$

The + or – sign is chosen with probability 0.5. δ is computed from a distribution, which prefers small values. This is realized as follows

$$\delta = \sum_{i=0}^{15} \alpha_i 2^i \qquad \alpha_i \in 0,1 \quad (11.4)$$

Before mutation we set $\alpha_i=0$. Then each α_i is mutated to 1 with probability $p_\delta=1/16$. Only $\alpha_i=1$ contributes to the sum. On the average there will be just one α_i with value 1, say α_j. Then δ is given by

$$\delta = 2^{-j} \quad (11.5)$$

The standard BGA mutation operator is able to generate any point in the hypercube with center x defined by $x_i \pm range_i$. But it generates values much more often in the neighborhood of x. In the above standard setting, the mutation operator is able to locate the optimal x_i up to a precision of $ramge_i \cdot 2^{-150}$.

As we mentioned before, we also solved the system identification problem with a LMS algorithm with the purpose of have a good reference mark. To monitor the convergence rate of the LMS algorithm, we computed a short term average of the squared error $e^2(n)$ using

$$ASE(m) = \frac{1}{K} \sum_{k=n+1}^{n+K} e^2(k) \tag{11.6}$$

where $m=n/K=1,2,....$ The averaging interval K may be selected to be (approximately) K=10N. The effect of the choice of the step size parameter Δ on the convergence rate of LMS algorithm may be observed by monitoring the ASE(m).

11.3.1 Genetic Algorithm for Optimization

The proposed genetic algorithm is as follows:

1. We use real numbers as a genetic representation of the problem.
2. We initialize variable i with zero (i=0).
3. We create an initial random population P_i, in this case (P_0). Each individual of the population has n dimensions and, each coefficient of the fuzzy system corresponds to one dimension. Since we are using a fuzzy system of 25 coefficients, then our search space is of n=25. The generated individuals have their coefficients in [-3,3].
4. We calculate the normalized fitness of each individual of the population using linear scaling with displacement (Mitchell, 1996), in the following form:

$$f_i^{'} = f_i + \frac{1}{N}\sum|f_i| + \left|\min_i(f_i)\right| \quad \forall i$$

5. We normalize the fitness of each individual using:

$$F_i = \frac{f_i^{'}}{\sum_{i=1}^{N} f_i^{'}} \quad \forall i$$

6. We sort the individuals from greater to lower fitness.
7. We use the truncated selection method, selecting the %T best individuals, for example if there are 500 individuals and, then we select 0.30*500=150 individuals.
8. We apply random crossover, to the individuals in the population (the 150 best ones) with the goal of creating a new population (of 500 individuals). Crossover with it self is not allowed, and all the individuals have to participate. To perform this operation we apply the genetic operator of extended intermediate recombination as follows:

9. If $x=(x_1,...,x_n)$ and $y=(y_1,...,y_n)$ are the parents, then the successors $z=(z_1,...,z_n)$ are calculated by, $z_i=x_i+\alpha_i(y_i-x_i)$ *for* $i=1,...,n$ where α is a scaling factor selected randomly in the interval $[-d,1+d]$. In intermediate recombination $d=0$, and for extended $d>0$, a good choice is $d=0.25$, which is the one that we used.
10. We apply the mutation genetic operator of BGA. In this case, we select an individual with probability $p_m=1/n$ (where n represents the working dimension, in this case n=25, which is the number of coefficients in the membership functions). The mutation operator calculates the new individuals z_i of the population in the following form: $z_i=x_i \pm range_i\ \delta$ we can note from this equation that we are actually adding to the original individual a value in the interval: $[-range_i, range_i]$ the range is defined as the search interval, which in this case is the domain of variable x_i, the sign ± is selected randomly with probability of 0.5, and is calculated using the following formula,

$$\delta = \sum_{i=0}^{m-1} \alpha_i 2^{-i} \qquad \alpha_i \in 0,1$$

11. Common used values in this equation are $m=16$ y $m=20$. Before mutation we initiate with $\alpha_i=0$, then for each α_i we mutate to 1 with probability $p_\delta=1/m$.
12. 10. Let i=i+1, and continue with step 4.

11.4 Automated Electrical Tuning of Television Sets

The problem of automated electrical tuning of television sets during production can be viewed as a controlling problem because we basically need to adjust the parameters to achieve the best image possible. We need to control the voltage, current intensity, and time to achieve the best image possible. We show in Figure 11.9(a) the frequency spectra for the imaging system of a typical television set. We also show in Figure 11.9(b) the acceptance region in the frequency spectra. This region is the one used by the human experts to decide on the appropriate quality of the image given by the television set.

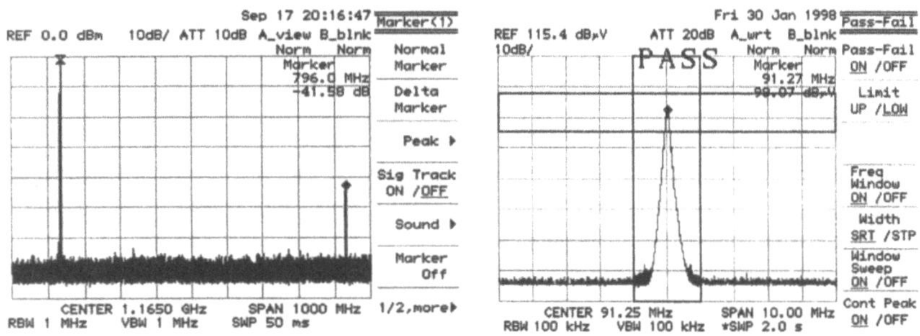

(a) (b)

Figure 11.9 (a) Frequency spectra for a typical television set, (b) acceptance region is shown.

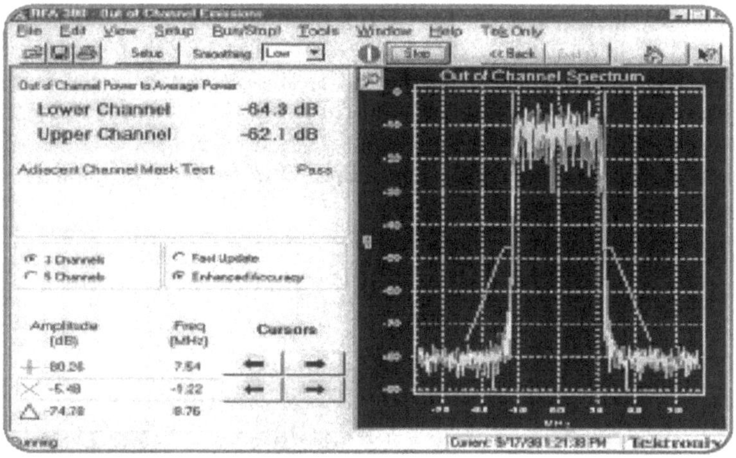

Figure 11.10 Actual output of the system showing a frequency spectrum.

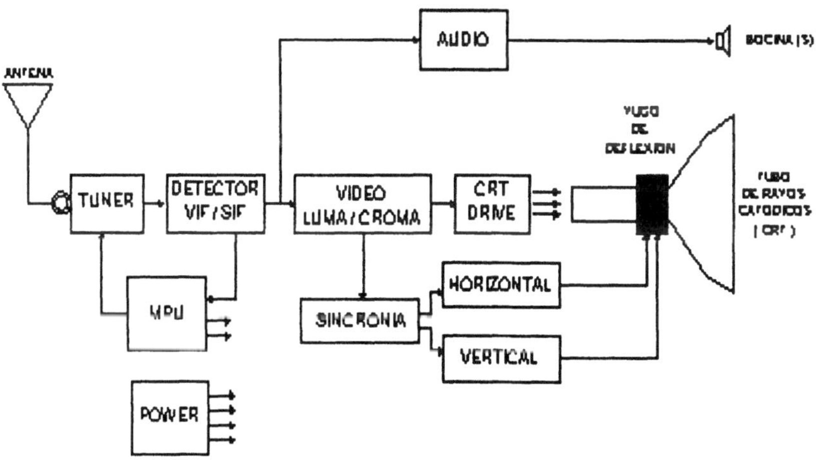

Figure 11.11 Schematics of the television set components.

We show in Figure 11.12 the traditional way of controlling the tuning process of television sets as done by the human experts in the plant. In this case, the human expert uses the output of the oscilloscope and the multi-meter to adjust the voltage and current by remote control. We show in Figure 11.13 the proposed intelligent system for controlling the electrical tuning process of television sets during manufacturing. We now have data acquisition cards to directly read the data from the oscilloscope and multi-meter, and also a computer program (intelligent system) in the computer to automatically adjust the image. For this purpose, the intelligent system has a knowledge base consisting of fuzzy rules, which contain the knowledge of the human experts in the tuning process. With the architecture shown in Figure 11.13 we really achieve automated control of the electrical tuning of television sets during production. The intelligent system was implemented in the MATLAB programming language, using the fuzzy logic toolbox of this language. The intelligent system works in real time in a real plant that produces television sets.

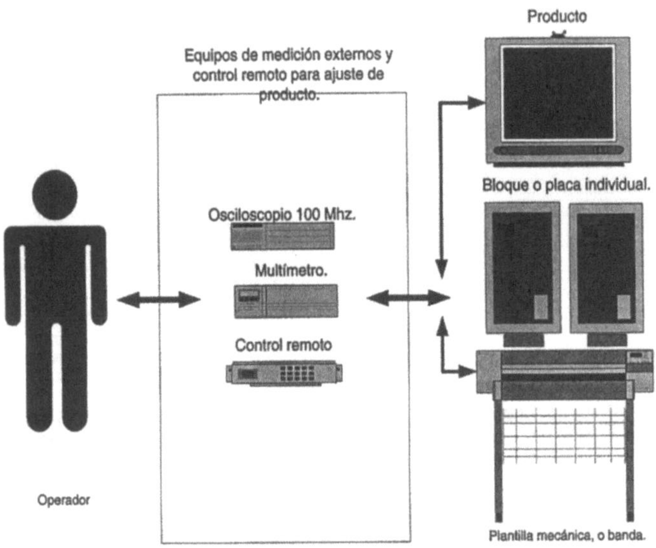

Figure 11.12 Traditional way of controlling the tuning process in television sets.

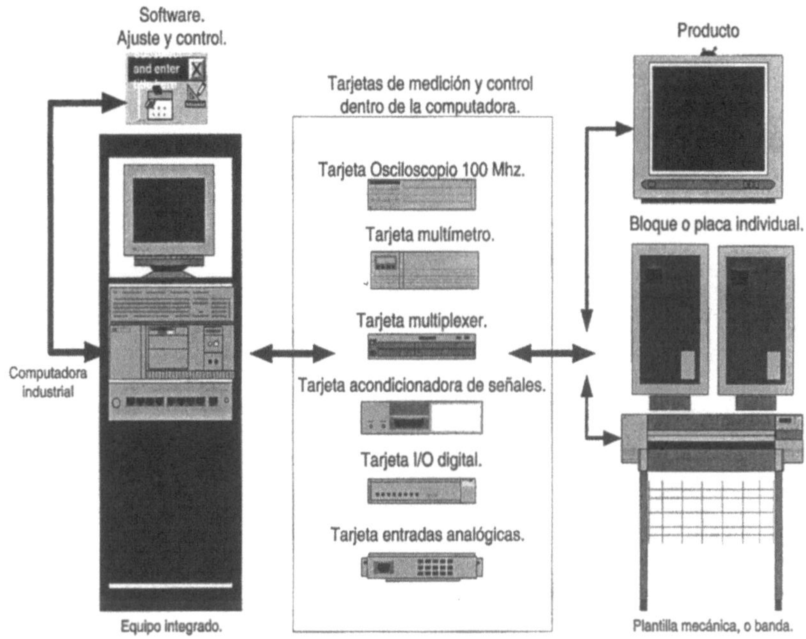

Figure 11.13 New approach for controlling the tuning process in television sets.

11.5 Intelligent System for Control

We developed an intelligent system for controlling the electrical tuning process during the television set manufacturing. We use fuzzy logic techniques (Zadeh, 1975) to automate the tuning process for television sets. We use a fuzzy system for controlling the voltage, current intensity, and frequency during the tuning process, so that the best possible quality of image is achieved. We also used genetic algorithms for optimizing the number of rules and parameters of the fuzzy system. We implemented the intelligent system for control in the MATLAB programming language (Nakamura, 1997). The fuzzy rule base was implemented in the fuzzy logic toolbox of MATLAB. On the other hand, we developed a computer program also in MATLAB for a breeder genetic algorithm that was used for optimizing the number of rules and parameters of the fuzzy system. The hybrid approach to the problem of control is a good choice, in this case, because of the complexity of the domain (Melin & Castillo, 2002).

We show in Figure 11.14 the non-linear surface of the intelligent system representing the model for controlling the electrical tuning process. We show in

Figure 11.15 the (Gaussian) membership functions for the output variable of the fuzzy system for controlling the electrical tuning process in the television set manufacturing. We show in Figure 11.16 the fuzzy rule viewer of the toolbox showing the use of the fuzzy system for specific values of voltage and current intensity. We show in Figure 11.17 the implementation of the fuzzy rule base for control in the MATLAB fuzzy logic toolbox. Finally, we show in Figure 11.18 the performance of the breeder genetic algorithm in optimizing the fuzzy system.

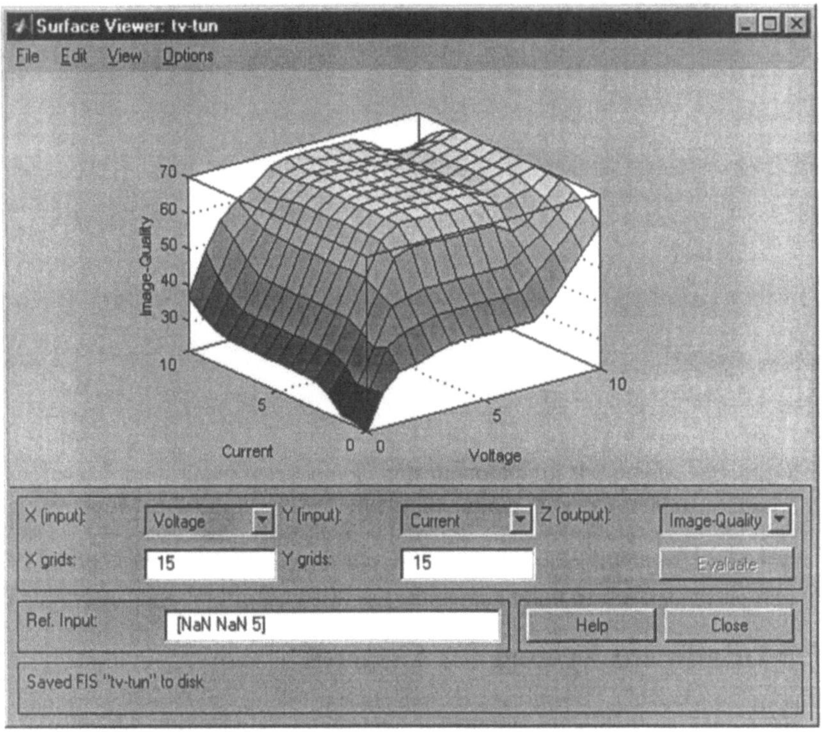

Figure 11.14 Non-linear surface for controlling the electrical tuning process in television sets.

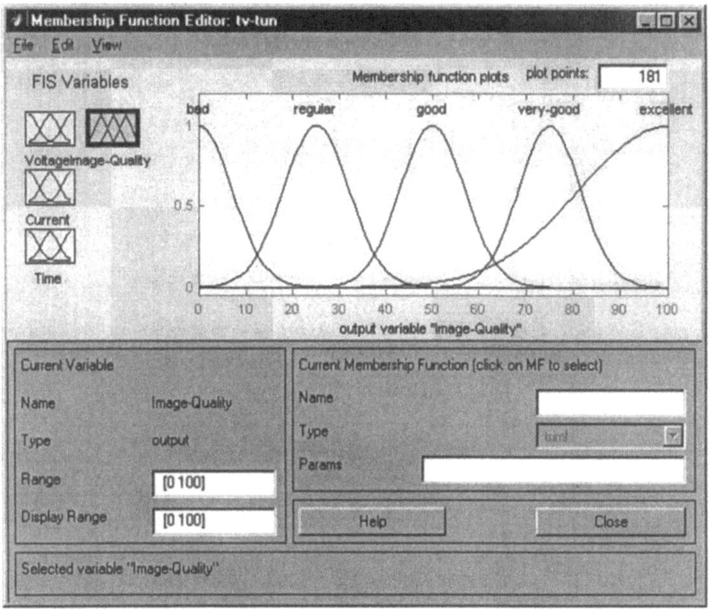

Figure 11.15 Membership functions for the output variable of the fuzzy system.

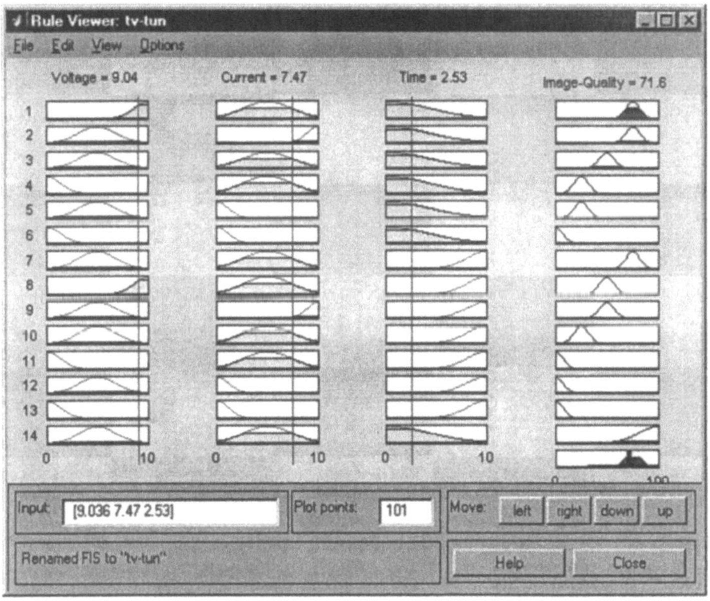

Figure 11.16 Fuzzy rule viewer showing the use of the intelligent system for specific values.

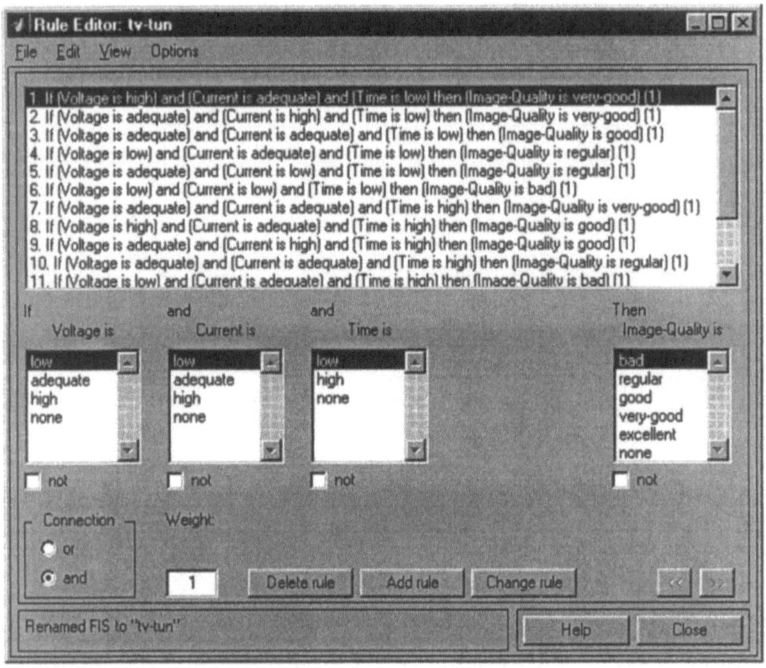

Figure 11.17 Implementation of the fuzzy rule base in the MATLAB fuzzy logic toolbox.

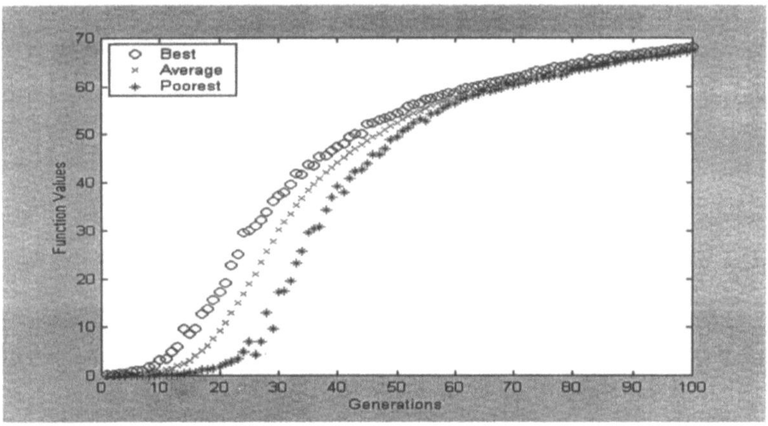

Figure 11.18 Results of optimizing the intelligent system with a genetic algorithm (%T=0.30, δ=16, best fitness =68.08 using 80 samples to calculate ASE).

11.6 Simulation Results

We did compare the performance of the intelligent system for control using the fuzzy-genetic approach with a system using a neuro-fuzzy approach, and a system with fuzzy logic alone. We used as a measure of efficiency for all the methods the percentage of times the desired quality of image was achieved. For this, we tested the systems for 30 days during production time. The results are shown in Table 11.2.

Table 11.2 Comparison of the methods for controlling the tuning process.

Method for Control	Efficiency
Manual Control	90%
Fuzzy Control	92.5%
Neuro-Fuzzy Control	95.5%
Fuzzy-Genetic Control	99.1%

From Table 11.2 we can see clearly how the use of soft computing techniques improves the quality of image in the television sets produced. The use of fuzzy logic alone helps in improving the efficiency. The use of neural networks helps in adapting the fuzzy system (Miller, Sutton & Werbos, 1995). Finally, the use of genetic algorithm improves even more the performance of the fuzzy system because not only the parameters are optimized also the rules are evolved.

11.7 Summary

We described in this chapter an intelligent system for controlling the electrical tuning process during the television set manufacturing. We used fuzzy logic to automate the tuning process for the television sets. We used the fuzzy system for controlling the voltage, current intensity, and frequency during the tuning process, so that the best possible quality of image is achieved. We also used a breeder genetic algorithm for optimizing the number of rules and parameters of the fuzzy system for control. We implemented the intelligent system for control in MATLAB with excellent simulation results. The intelligent system was also tested in a real production plant with excellent results.

Chapter 12

Intelligent Manufacturing of Batteries

We describe in this chapter, different hybrid intelligent approaches for controlling non-linear dynamical systems in manufacturing applications. The hybrid approaches combine soft computing techniques and mathematical models to achieve the goal of controlling the manufacturing process to follow a desired production plan. We have developed several hybrid architectures that combine fuzzy logic, neural networks, and genetic algorithms, to compare the performance of each of these combinations and decide on the best one for our purpose. We consider the case of controlling non-linear electrochemical processes to test our hybrid approach for control. Electrochemical processes, like the ones used in battery formation, are very complex and for this reason very difficult to control. We have achieved very good results using fuzzy logic for control, neural networks for modeling the process, and genetic algorithms for tuning the hybrid intelligent system.

This chapter also shows a fuzzy logic system for the formation of batteries using a high scale integration device such as the FLASH PSD813F1, which is a micro-controller. It includes a CPLD (COMPLEX PLD) embedded, which allows implementing logical functions defined by the user in its 73 inputs 19 outputs AND-OR array. The fuzzy logic algorithm is summarized in a look up table, described by Boolean equations implemented in the PSD813F1 CPLD.

This chapter also describes an embedded fuzzy control system for a quality control test in the manufacturing of power batteries. This battery test consists in discharging a battery for a certain time at a constant current value, all depending on the battery model. This system is able to improve on others, as it minimizes the problems presented when the test is initiated and when there are false contacts during the test, protecting the battery itself and test equipment. At the moment, a proportional controller is used that even it has demonstrated to be efficient, it has the disadvantages mentioned above.

In this fuzzy logic system for the quality control of batteries, we use again a high scale integration device such as the FLASH PSD813F1, which is a micro-controller. It includes a CPLD (COMPLEX PLD) embedded, which allows implementing logical functions defined by the user in its 73 inputs 19 outputs AND-OR array. The fuzzy logic algorithm is summarized in a look up table, described by Boolean equations implemented in the PSD813F1 CPLD.

12.1 Intelligent Control of the Battery Charging Process

We describe in this section a hybrid intelligent approach for adaptive control of non-linear dynamical systems. We propose a hybrid approach to control dynamical systems that combines neural networks, fuzzy logic and genetic algorithms. We illustrate this approach with the problem of controlling a non-linear electrochemical process in a real world plant. The dynamics of an electrochemical system is non-linear and for this reason can be very difficult to predict in an accurate manner (Melin & Castillo, 1999). Also, mathematical models of electrochemical processes are difficult to derive and they are not very accurate. Traditionally, the models that have been used for electrochemical processes are from Statistics, but these models do not approximate the dynamic behavior of the processes with the accuracy required in practice. We need adaptive control of the electrochemical process to achieve on-line control of the production line. Of course, adaptive control is easier to achieve if one uses a reference model of the process (Melin & Castillo, 2002). The problem is how to obtain a good model for the process, considering that mathematical models are very inefficient. The use of soft computing techniques has been recognized as a different approach to modeling with good results, and for this reason we decided to turn to these techniques for our problem (Castillo & Melin, 2001).

We have tested fuzzy logic and neural networks for modeling the electrochemical processes and we have decided that the best choice for this case is to use neural networks. The other part of the problem is how to control the non-linear electrochemical process in the desired way to achieve the production with the required quality. We tested different techniques and we arrived to the conclusion that fuzzy logic (Zadeh, 1975) was the best choice for this case. We developed a set of fuzzy rules with the expert knowledge for controlling the manufacturing process. The membership functions for the linguistic variables in the rules were tuned using a specific genetic algorithm. The genetic algorithm was used for searching the parameter space of the membership functions using real data from production lines. Our particular neuro-fuzzy-genetic approach has been implemented as an intelligent system to control the formation of batteries in a real plant with very good results.

12.1.1. Problem Description

In a battery a process of conversion of chemical energy into electrical energy is carried out. The chemical energy contained in the electrode and electrolyte is converted into electrical power by means of electrochemical reactions. When connecting the battery to a source of direct current a flow of electrons takes place for the external circuit, and of ions inside the battery, giving an accumulation of load in the battery. The quantity of electric current that is required to load the battery is determined by an unalterable law of nature, that was postulated by Michael Faraday, which is known as the Law of Faraday (Bode, Brodd, & Kordesch, 1977). Faraday found that the quantity of electric power required to perform an electrochemical change in a metal is related to the relative weight of the metal. In the specific case of lead this is considered to be 260 amperes hour per kilogram of positive active material for cell. In practice, more energy is required to counteract the losses due to the heat.

We show in Table 12.1 experimental data for a specific type of battery with different sizes of the plates, and different number of plates for each cell. In this table, we show the load time and the average current needed for the respective load. In Table 12.1 we can observe that to form a battery we need to apply a particular current intensity during a certain amount of time to achieve the required loading for the battery. For example, from the first line of Table12.1 we can see that a battery with positive plate of 1.524 mm and negative plate of 1.270 mm requires 155 amperes during 1 hour or 2.2 amperes during 72 hours.

Table 12.1 Experimental data for different types of batteries

Type of Plate						
	Positive 1.524 mm Negative 1.270 mm			Positive 1.778 mm Negative 1.524 mm		
Plate cell	Total A. H.	72 hr Amp	96 hr Amp	Total A.H.	72 hr Amp	96 hr Amp
7	155	2.2	1.6	165	2.4	1.8
9	180	2.8	2.0	200	2.8	2.2
11	230	3.2	2.4	245	3.4	2.4
13	260	3.6	2.6	295	4.0	3.0
15	300	4.2	3.0	345	4.8	3.6
17	400	5.6	4.2	415	5.8	4.4

The goal of the manufacturers of batteries is to reduce the time required to load the battery. However, current intensity can't be increased arbitrarily because of the physical characteristics of the specific battery. If the current is increased too much, the temperature in the battery will go over a safe temperature value eventually causing the destruction of the battery. The problem is then, of finding out how much can we increase the current without causing the battery to explode. Of course, we need to control the current during the process of battery formation, and we want to do this in such a way as minimize the time of loading.

With the purpose of finding out, during the process of formation of the battery, the appropriate values of current intensity without surpassing the limits of temperature (Hehner, & Orsino, 1985), we propose three systems for intelligent control of the process. The first one uses only fuzzy logic for control and statistical models of the process. The second one uses a neuro-fuzzy approach to develop the fuzzy rules controlling the electrochemical process. The third approach uses neural networks to model the process, fuzzy rules for control and genetic algorithms to tune the membership functions.

12.1.2 Fuzzy Method for Control

In this approach we use a statistical model to represent the electrochemical process and a fuzzy rule base for process control. The temperature in the battery depends on the electrical current that circulates in it during its formation, this means that to maintain the temperature below a specific threshold it is important to control the intensity of the current. Therefore for this case the independent variable is the average current I, and the dependent variable is the average temperature T. A simple statistical model can be stated as follows: $T = \beta_o + \beta_1 I$ where β_o and β_1 are parameters to be estimated (by least squares) using real data for this problem. In Table 12.2, we show experimental values for a battery of 6 Volts, which according to manufacturer's specifications should be loaded by using 200 amperes hour.

Using the data from Table 12.2 we can obtain (by the least squares method) the values of β_o and β_1. The equations is as follows:

$$T = 88.03 + 2.5304\, I \qquad (12.1)$$

with correlation value of only 0.57 which is understandable because of the complexity of the data. For the fuzzy controller we used as input variables, the temperature T and the change of temperature dT/dt, and as output variable the current intensity that should be applied to the battery. In Figure 12.1 we show the architecture of our control system.

Table 12.2 Experimental values for a battery of 6 Volts.

Hrs	T	I	Hrs	T	I
21:00	111	5.22	23:00	93	3.53
23:00	100	5.21	1:00	91	3.40
1:00	105	5.52	3:00	92	3.32
3:00	100	5.66	5:00	96	3.16
5:00	100	5.60	7:00	98	3.10
7:00	97	5.72	9:00	98	3.14
9:00	92	4.82	11:00	102	3.12
11:00	95	4.32	13:00	99	3.03
13:00	102	4.10	15:00	98	3.05
15:00	103	4.05	17:00	97	3.06
17:00	100	3.40	19:00	95	2.96
19:00	97	3.77	21:00	94	2.60
21:00	94	3.62	23:00	96	2.76

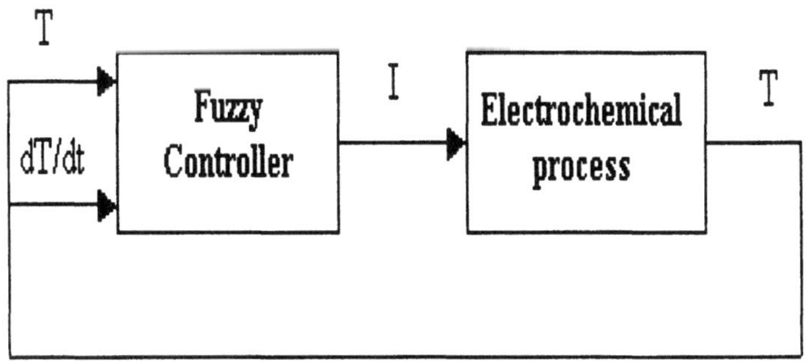

Figure 12.1 Fuzzy Control of the process

The control method was implemented in the MATLAB programming language (Nakamura, 1997). For each of the linguistic variables it was considered convenient to use five linguistic terms. We have 25 fuzzy rules because we are using 5 linguistic terms for each variable.

12.1.3 Neuro-Fuzzy Method for Control

Since it is difficult to tune a particular fuzzy inference system to model a complex dynamical system (Melin & Castillo, 2002) it is convenient to use adaptive fuzzy inference systems. Adaptive neuro-fuzzy inference systems (ANFIS) can be used to adapt the membership functions and consequents of the rule base according to historical data of the problem (Jang, Sun & Mizutani, 1997). In this case, we can use the data from Table 12.2 and apply the ANFIS methodology to find the best fuzzy system for our problem. We used the fuzzy logic toolbox of MATLAB to apply the ANFIS methodology to our problem with 5 membership functions and first order Sugeno functions in the consequents.

In Figure 12.2 we show the architecture of the control system. The fuzzy rule base was implemented in the MATLAB programming language. We show in Figure 12.3 the non-linear surface for control. We only used 5 rules because there is only one input variable (temperature) and one output variable (electrical current) with 5 linguistic values each.

We show in Figure 12.4 the results of training with the ANFIS methodology using the data of Table 12.2. We can see from this figure that the method approximates very well real data for the problem.

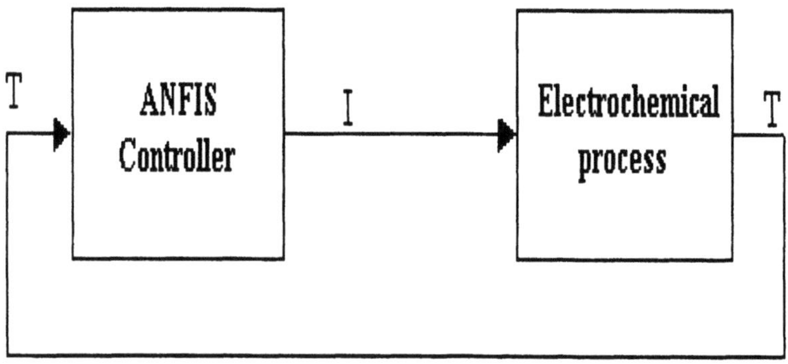

Figure 12.2 ANFIS Control of the process.

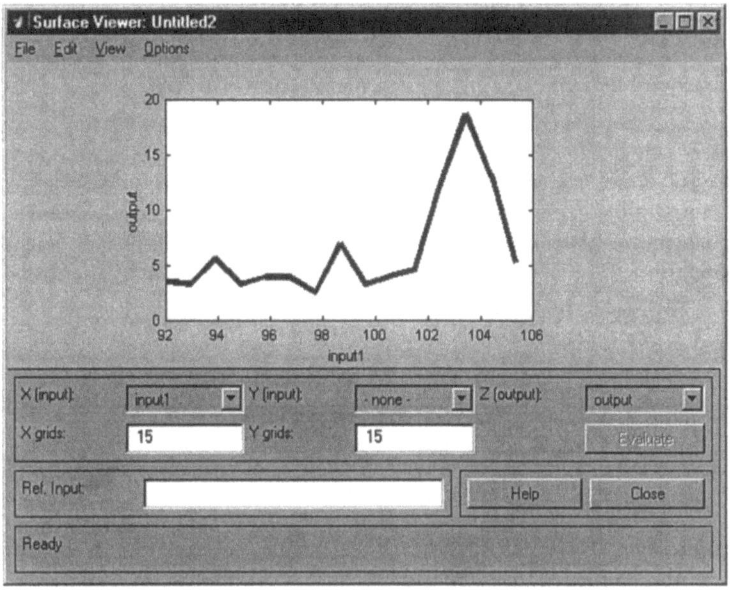

Figure 12.3 ANFIS non-linear surface for the process.

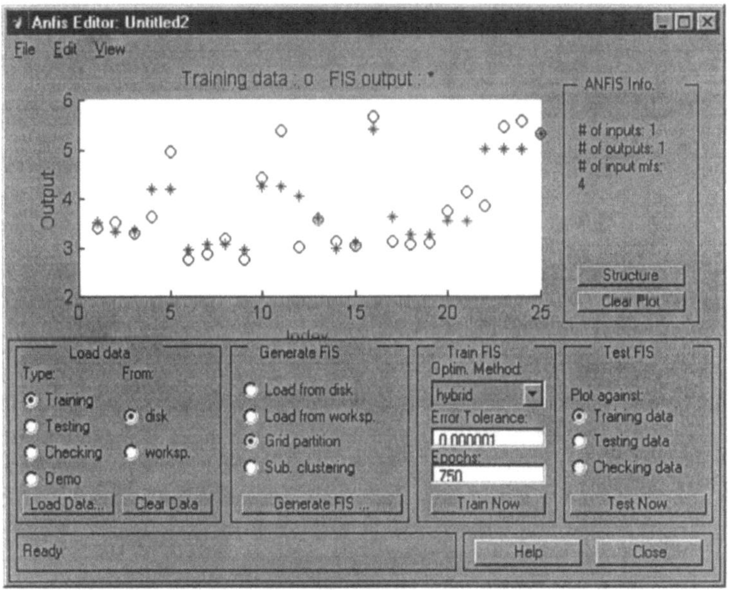

Figure 12.4 ANFIS training with the data from Table 2.

12.1.4 Neuro-Fuzzy-Genetic Method for Control

In this case, we consider using neural networks (Miller, Sutton & Werbos, 1995) for modeling the electrochemical process, fuzzy logic for controlling the electrical current and genetic algorithms (Man, Tang, & Kwong, 1999) for adapting the membership functions of the fuzzy system. The general architecture of the control system is shown in Figure 12.5.

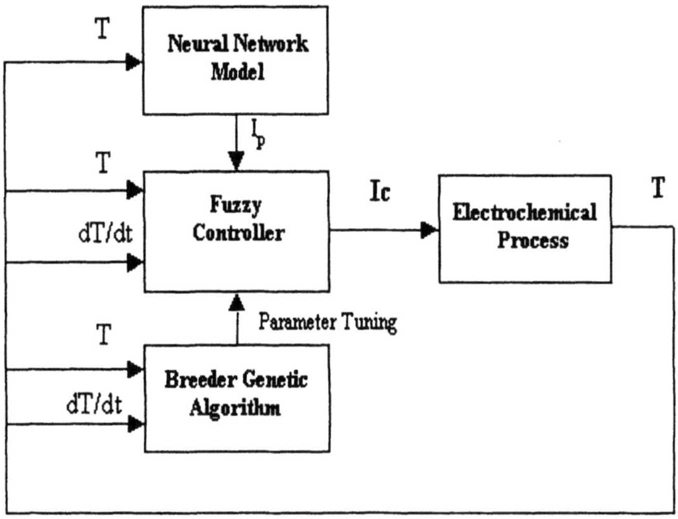

Figure 12.5 Neuro-Fuzzy-Genetic Control

A multi-layer feed-forward neural network was used for modeling the electrochemical process. We used the data form Table 12.2 and the Levenberg-Marquardt learning algorithm to train the neural network. We used a three layer neural network with 15 nodes in the hidden layer. The results of training for 2000 epochs are shown in Figure 12.6. The sum of squared errors was reduced from about 200 initially to 11.25 at the end, which is a very good approximation in this case.

We show in Figure 12.7 the results of applying the genetic algorithm, described before, to the problem of optimizing the hybrid intelligent system. We can see in this figure that the fitness of the intelligent system for control is optimized. The best value of fitness for the hybrid system was of 68.08, as shown in Figure 12.7.

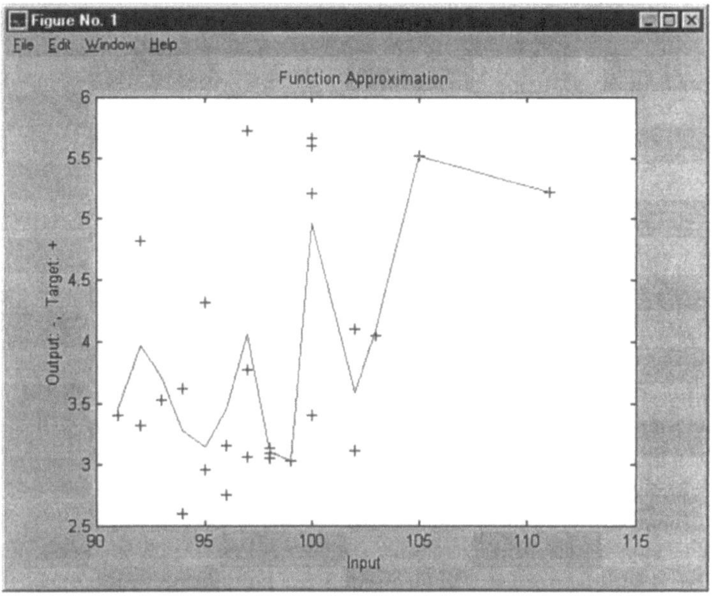

Figure 12.6 Final training for the neural network.

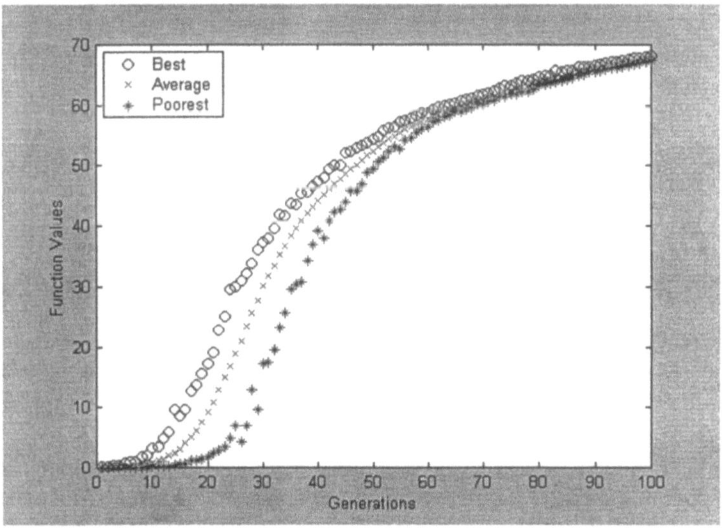

Figure 12.7 Results of optimizing the intelligent system with a genetic algorithm (%T=0.30, δ=16, best fitness =68.08 using 80 samples to calculate ASE).

The fuzzy rule base was implemented in the Fuzzy Logic Toolbox of MATLAB. We used 25 fuzzy rules because there were 5 linguistic terms for each input variable. The initial membership functions were obtained from experts, but they were improved using a specific genetic algorithm to arrive to the final functions.

12.2 Hardware Implementation of the Fuzzy Controller for the Charging Process

There exists an imperious need of shortening the time of formation of the batteries at the production line, this is due to the great demand of this product. In this section, we describe a novel technique in this area to optimize the time of manufacturing. This technique uses a fuzzy approach (implemented in hardware) to increase the ampere rate in order to minimize the formation time.

This section shows a fuzzy logic system for the formation of batteries using a high scale integration device, the FLASH PSD813F1, which is a micro-controller supporting the hardware implementation. It includes a CPLD (COMPLEX PLD) embedded, it allows implementing logical functions defined by the user in its 73 inputs 19 outputs AND-OR array. The fuzzy logic algorithm is summarized in a look up table, described by Boolean equations implemented in the PSD813F1 CPLD.

The fuzzy control for the formation of batteries using the CPLD of the PSD813F1, basically controls the current intensity applied to the battery depending on its temperature and the change of temperature. This fuzzy control system allows the automation of the process and reduces the time of formation, which actually needs human supervision.

12.2.1 Introduction

In a battery it is carried out a process of conversion of chemical energy contained in the active materials, as they are the electrodes and the electrolyte, to electric power by means of electrochemical reactions. When a battery is connected to a current source, a flow of electrons takes place for the external circuit, and of ions inside the battery, giving an accumulation of charges as a result in each terminal of the battery and in consequence a battery loaded. This process is denominated formation of a battery.

The amount of electrical current that is required to form the plates of the battery is determined by an unalterable law of nature that was postulated by the English Michael Faraday and that one knows as the Law of Faraday. Faraday found that the quantity of electrical energy required to carry out an electrochemical change in a metal like the lead, is related with the relative weight of such a metal. In practice, more energy is required to conteract the losses due to the heat and to the generation of gas. However, many plate formations are accomplished using 125 ampere hours per pound of positive active material per cell (Hehner, 1981).

There exists an imperious need to shortening the time of formation of the batteries at the production line, this is due to the great demand of this product. In this section we are proposing a novel technique in this area to optimize the time of manufacturing. This technique uses a fuzzy approach to increase the ampere rate in order to minimize the formation time. The problem is that the current increase can not be arbitrary, it depends on the capacity of the battery model, and the increase of temperature that is generated inside it. An inappropriate ampere rate to forming a battery can damaged and even destroyed it (Sepúlveda, Castillo, Montiel, & Lopez, 1998).

In this work the fuzzy system for the formation of batteries, that was implemented in the CPLD of the PSD813F1 (Harris & Rootz, 1999), basically controls the current intensity applied to the battery depending on the temperature and the change of temperature. This fuzzy control system allows the automation of the process and reduces the time of formation, which actually needs human supervision.

12.2.2 Fuzzy Controller

The method described here to implement the fuzzy controller in the CPLD consists of logically synthesizing the controller into a set of Boolean equations (Manzoul 1994), then reduce it to a look up table. After the knowledge base and the set of all possible inputs are known, the look up table depends on the number of inputs of the input variables. More precisely the number of rows is equal to the product of the number of singletons of each input (Manzoul & Jayabharathi, 1991). The look up table is described by a set of Boolean equations, which is an acceptable input to the CPLD.

To realize the fuzzy controller on a CPLD, first of all, we have to reduce the fuzzy controller to a set of Boolean equations and then these equations can be implemented on the CPLD.

For a two-input single-output fuzzy controller, the IF-THEN rule is written as:

IF x is A_i AND y is B_i THEN z is C_i

Here x and A_i are defined in the universe of discourse A^I, y and B_i are defined in the input universe of discourse B^I, while z and C_i are defined in the output universe of discourse C^I.

This fuzzy rule can be put into the simpler form

$$\mathbf{A \times B \rightarrow C}$$

this rule can be transformed into a ternary fuzzy relation *Rm* based on Mamdani's fuzzy implication function (Jang, Sun & Mizutani, 1997):

$$R_m(A,B,C) = (AxB)xC =$$
$$= \int_{XxYxZ} \mu_A(x) \wedge \mu_B(x) \wedge \mu_C(x)/(x,y,z).$$

where the resulting C' is expressed as

$$C' = (A'xB') \circ (AxB \to C).$$

which can be decomposed as (Jang, Sun & Mizutani, 1997):

$$C' = [A' \circ (A \to C)] \cap [B' \circ (B \to C)]$$

The preceding expression states that the resulting consequence C' can be expressed as the intersection of

$$C'_1 = A' \circ (A \to C)$$

and

$$C'_2 = B' \circ (B \to C)$$

each of one corresponds to a single-input single-output fuzzy rule as

IF $\boldsymbol{x}$ is $\boldsymbol{A_i}$ THEN $\boldsymbol{y}$ is $\boldsymbol{B_i}$

where x and A_i are defined in the universe of discourse A^I, while y and B_i are defined in the output universe of discourse B^I.

If the dimensions of the two universes of discourse A^I and B^I are:

$$\dim[\boldsymbol{A^I}] = \boldsymbol{q}$$
$$\dim[\boldsymbol{B^I}] = \boldsymbol{p}$$

where $\boldsymbol{q}$ is the number of singletons in the input universe of discourse, and $\boldsymbol{p}$ is the number of singletons in the output universe of discourse, then the fuzzy relation of the system is computed as (Zadeh, 1973),

$$\boldsymbol{R} = \vee (\boldsymbol{A_i} \ \boldsymbol{B_i})$$

where $\vee$ is the ***max*** operator, the dimension of the relation is:

$$\dim[\boldsymbol{R}^1] = \boldsymbol{q} \times \boldsymbol{p}$$

So given any input $\boldsymbol{x}$, the fuzzy output $\boldsymbol{B}$ is computed using the compositional rule of inference (Tsoukalas & Uhrig, 1997),

$$\mathbf{y} = \boldsymbol{x} \circ \boldsymbol{R}$$

where $\circ$ is the max-min composition of the fuzzy relation.

So for the two-input single output

$$C' = C'_1 \cap C'_2$$

we have for C'_1

$$\boldsymbol{R'}_1 = \vee (\boldsymbol{A}_i \;\; \boldsymbol{C}_i)$$

and for C'_2

$$\boldsymbol{R'}_2 = \vee (\boldsymbol{B}_i \;\; \boldsymbol{C}_i)$$

For any inputs x, y, the fuzzy output C' is computed using the intersection of the compositional rules as:

$$C' = (\boldsymbol{x} \circ \boldsymbol{R'}_1) \cap (\mathbf{y} \circ \boldsymbol{R'}_2)$$

To achieve a good control in the formation of the batteries we consider a fuzzy controller. The system can be very complex in a real situation, but for space reasons we describe it with the following linguistic variables:

input: Temperature, with three membership functions: Negative Big NB, Normal NORMAL and Positive Big PB.

input: ΔTemperature, with three membership functions: Negative Big NB, Cero ZERO and Positive Big PB.

output: Current C, with three membership functions: Negative Big NB, Normal NORMAL y Positive Big PB.

The system is described by the following three rules:

1 IF Temp is NB AND ΔTemp is NB
THEN Current is PB

2 IF Temp is NORMAL AND ΔTemp is NORMAL THEN Current is NORMAL

3 IF Temp is PB AND ΔTemp is PB
THEN Current is NB

The original ranges for the inputs Temp and ΔTemp are: 90 to 110 °F and -5 to 5 °F respectively and 2 to 10 Ampere for the output Current C . The membership values are in the interval [0, 1], where 0 denotes no membership and 1 denotes full membership. Figure 12.8 shows the membership functions of the linguistic variables.

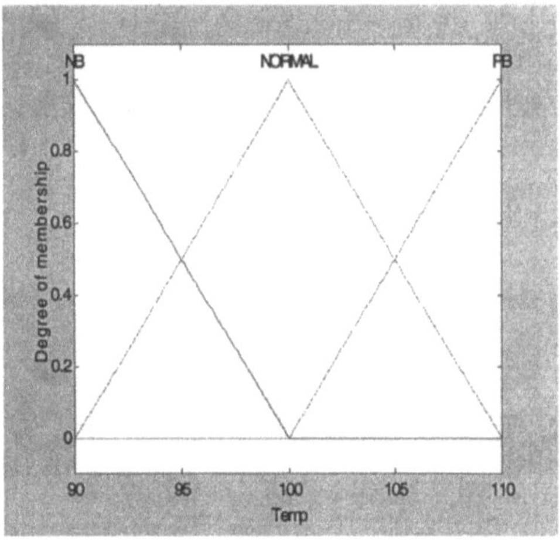

a)

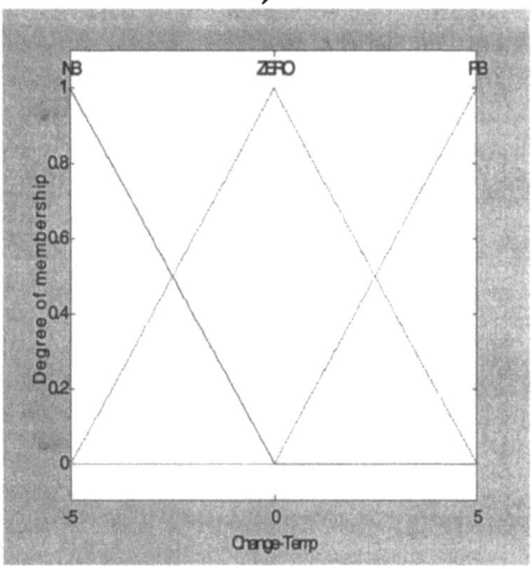

b)

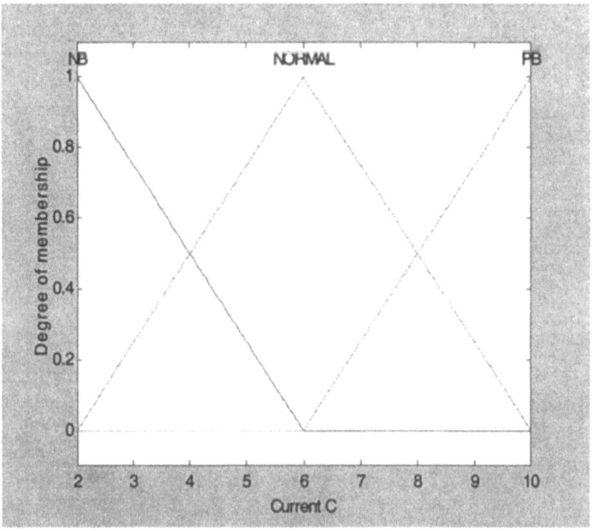

c)

Figure 12.8. Membership functions, a) input Temperature, Temp., b) input ΔTemp ,c) output Current C.

Assume dim[***A***] = dim[***B***] = dim[***C***]= 7, and the three rules of the system are expressed numerically as:

1. IF [1.0 .5 0 0 0] AND [1.0 .6 0 0 0]
   THEN [0 0 0 .5 1.0]

2. IF [0 .5 1.0 .5 0] AND [0 .3 1 .3 0]
   THEN [0 .5 1.0 .5 0]

3. IF [0 0 0 .5 1.0] AND [0 0 0 .6 1]
   THEN [1.0 .5 0 0 0]

Using the equation for the fuzzy relation $\boldsymbol{R'_1}$, we have:

$$R'_1 = \begin{bmatrix} 0 & 0 & 0 & .5 & 1 \\ 0 & .5 & .5 & .5 & .5 \\ 0 & .5 & 1 & .5 & 0 \\ .5 & .5 & .5 & .5 & 0 \\ 1 & .5 & 0 & 0 & 0 \end{bmatrix}$$

and for the fuzzy relation $\boldsymbol{R'_2}$,

$$R'_2 = \begin{bmatrix} 0 & 0 & 0 & .5 & 1 \\ 0 & .3 & .3 & .5 & .6 \\ 0 & .5 & 1 & .5 & 0 \\ .6 & .5 & .3 & .3 & 0 \\ 1 & .5 & 0 & 0 & 0 \end{bmatrix}$$

For example if the temperature of the battery is 94 °F, and the change of temperature is –4 °F that corresponds to the fuzzy values

93 = [.0 1.0 .0 .0 .0]

and

-4 = [.0 1.0 .0 .0 .0]

as can be seen on tables 12.3 and 12.4 respectively.

To determine the fuzzy output, we employed the intersection of the compositional rules of inference given on row A2B2 of table 12.5. The fuzzy output would be:

Fuzzy output = [.0 .3 .3 .5 .5]

After deffuzzification, the output is found to be, 7 Ampere. In table 12.5 we can see the complete computations of the controller, and the look up table representing the fuzzy controller is given in table 12.6.

Table 12.3 Summary of the controller computations for A' ∘ (A→C).

	Input (Temp)	Fuzzified input	Fuzzified output
A1	0 to 90	1.0 .0 .0 .0 .0	.0 .0 .0 .5 1.0
A2	90.1 to 95	.0 1.0 .0 .0 .0	.0 .5 .5 .5 .5
A3	95.1 to 100	.0 .0 1.0 .0 .0	.0 .5 1.0 .5 0
A4	100.1 to 105	.0 .0 .0 1.0 .0	.5 .5 .5 .5 .0
A5	105.1 to ∞	.0 .0 .0 .0 1.0	1.0 .5 .0 .0 .0

Table 12.4 Summary of the controller computations for B' ∘ (B→C).

	Input (ΔTemp)	Fuzzified input	Fuzzified output
B1	0 to -5	1.0 .0 .0 .0 .0	.0 .0 .0 .5 1.0
B2	-4.9 to -2.5	.0 1.0 .0 .0 .0	.0 .3 .3 .5 .6
B3	-2.49 to 0	.0 .0 1.0 .0 .0	.0 .5 1.0 .5 0
B4	0.1 to 2.5	.0 .0 .0 1.0 .0	.6 .5 .3 .3 .0
B5	2.51 to ∞	.0 .0 .0 .0 1.0	1.0 .5 .0 .0 .0

Table 12.5 Summary of the controller computations for:
C'= [A' ∘ (A→C)] ∩ [B' ∘ (B→C)]

	Fuzzified output	Output
A1B1	.0 .0 .0 .5 1.0	9
A1B2	.0 .0 .0 .5 .6	9
A1B3	.0 .0 .0 .5 .0	8
A1B4	.0 .0 .0 .3 .0	8
A1B5	.0 .0 .0 .0 .0	0
A2B1	.0 .0 .0 .5 .5	9
A2B2	.0 .3 .3 .5 .5	7
A2B3	.0 .5 .5 .5 .0	6
A2B4	.0 .5 .3 .3 .0	5
A2B5	.0 .5 .0 .0 .0	4
A3B1	.0 .0 .0 .5 .0	8
A3B2	.0 .3 .3 .5 .5	7
A3B3	.5 .5 .1 .5 .0	5
A3B4	.5 .5 .3 .3 .0	4
A3B5	.5 .5 .0 .0 .0	3
A4B1	.0 .0 .0 .5 .0	8
A4B2	.0 .3 .3 .5 .0	6
A4B3	.5 .5 .5 .5 .0	5
A4B4	.5 .5 .3 .3 .0	4
A4B5	.5 .5 .0 .0 .0	3
A5B1	.0 .0 .0 .0 .0	0
A5B2	.0 .3 .0 .0 .0	4
A5B3	.5 .5 .0 .0 .0	3
A5B4	.6 .5 .0 .0 .0	3
A5B5	1.0 .5 .0 .0 .0	2

Table 12.6 Look up table for the fuzzy controller.

Input	Output f3 f2 f1 f0	Output
A1B1	1 0 0 1	9
A1B2	1 0 0 1	9
A1B3	1 0 0 0	8
A1B4	1 0 0 0	8
A1B5	0 0 0 0	0
A2B1	1 0 0 0	9
A2B2	0 1 1 1	7
A2B3	0 1 1 0	6
A2B4	0 1 0 1	5
A2B5	0 1 0 0	4
A3B1	1 0 0 0	8
A3B2	0 1 1 1	7
A3B3	0 1 0 1	5
A3B4	0 1 0 0	4
A3B5	0 0 0 0	3
A4B1	1 0 0 0	8
A4B2	0 1 1 0	6
A4B3	0 1 0 1	5
A4B4	0 1 0 0	4
A4B5	0 0 1 1	3
A5B1	0 0 0 0	0
A5B2	0 1 0 0	4
A5B3	0 0 1 1	3
A5B4	0 0 1 1	3
A5B5	0 0 1 0	2

Finally, the look up table can be described by four variable Boolean equations, since the input universe of discourse and the output universe of discourse have dimensions equal to seven:

$f0 = \Sigma\ (1,2,7,9,12,13,18,20,23,24)$
$f1 = \Sigma\ (7,8,12,17,20,23,24)$
$f2 = \Sigma\ (7,8,9,10,12,13,14,17,18,19,22)$
$f3 = \Sigma\ (1,2,3,4,6,11,16,)$

To compare the output results of the fuzzy system proposed for the CPLD, it was designed in the MATLAB fuzzy toolbox, a fuzzy system with two inputs and one output as can be seen in Figure 12.9. We have three membership functions for input Temperature, NB (Negative Big) NORMAL and PB (Positive Big); three membership functions for input ΔTemperature, NB (Negative Big) ZERO and PB (Positive Big)as shown in Figure 12.8a.and 12.8b. For the output Current, there are also three membership functions, NB (Negative Big) NORMAL and PB (Positive Big) as shown in Figure 12.8c.

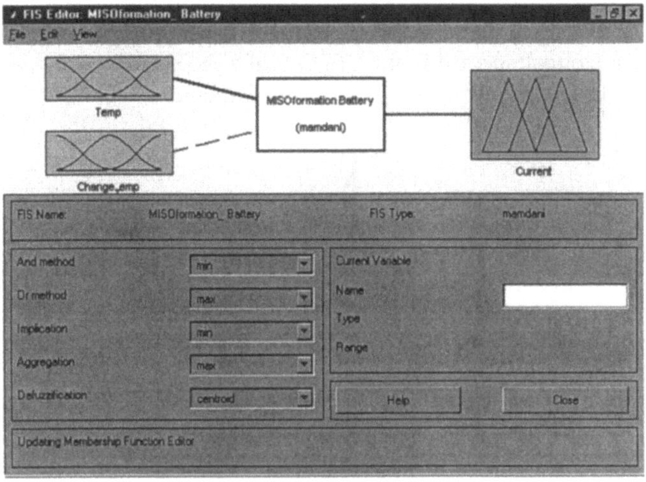

Figure 12.9 Two inputs, one output Fuzzy System.

In Figure 12.10 we can see the output value 7.21, which corresponds to 94 °F as input temperature, in the range 90.1 to 95 of the table 1,and to -4.°F as input of change of temperature in the range –4.9 to –2.5 of table 12.4.

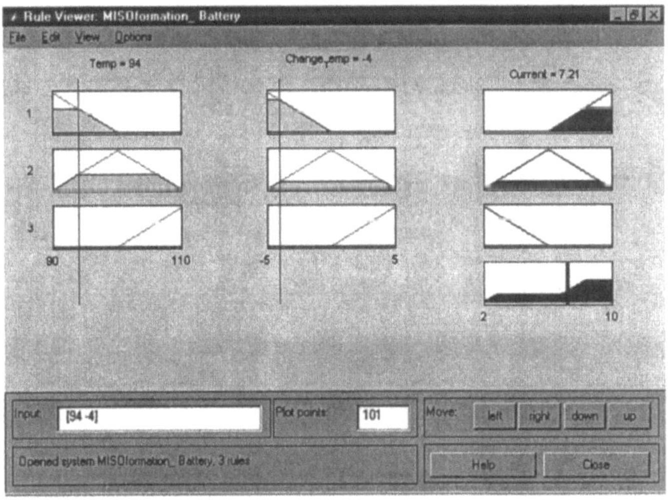

Figure 12.10 Summary of the fuzzy control for the inputs 94°F and -4°F , with an output=7.21 Amps.

12.2.3 Implementation of the Fuzzy Control

The Boolean equations obtained as a result of the logic synthesis of the fuzzy controller can be implemented in the CPLD of the PSD813F1. Figure 12.11 shows the diagram for the fuzzy control of the battery formation.

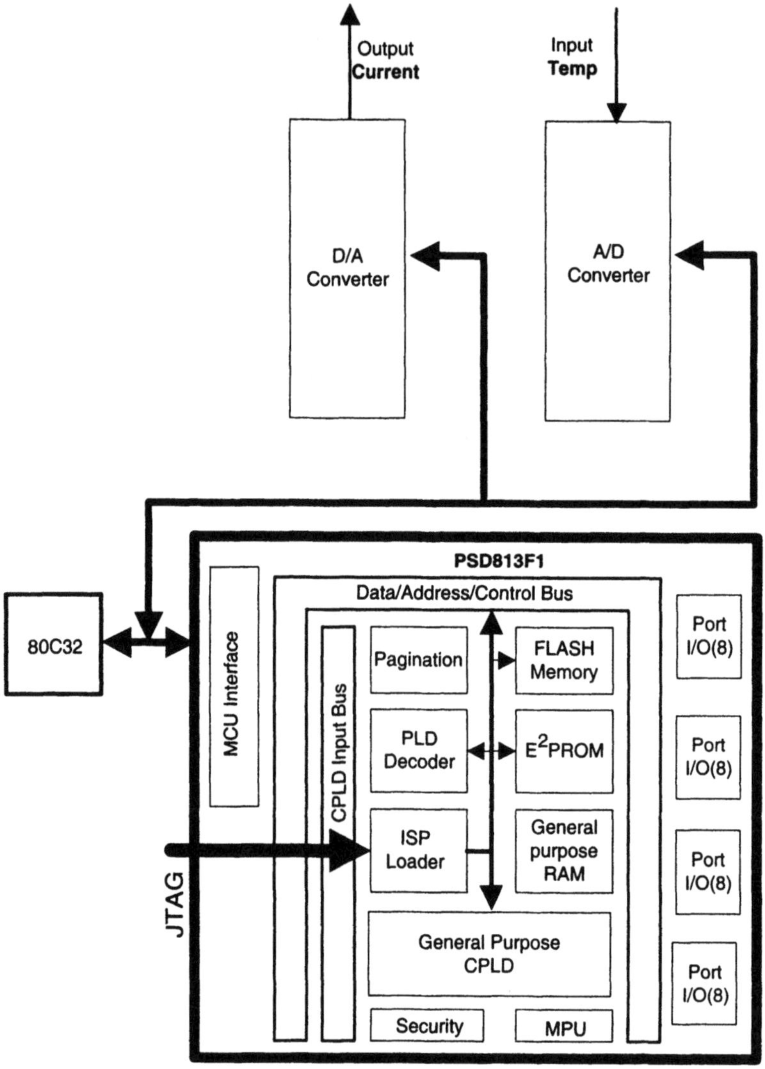

Figure 12.11 Fuzzy Controller.

The PSD813F1 family includes a JTAG channel for In-System Programming (ISP). This ISP function is an extension of the typical JTAG boundary-scan function. It is an implementation of the JTAG-ISC (In-System Configuration) specification that is becoming an industry standard. The PSD allows the user to define and configure the JTAG ISC interface to meet the specifications, and needs no assistance from the embedded microcontroller (MCU), instead it uses the JTAG channel and the PSDsoft, which is an integrated system development software tool.

To program the fuzzy controller, it is necessary to configure and program the PSD, through the PSDsoft. This tool runs on a PC in the Windows 95/98 and Windows NT environments. PSDsoft supports the configuration of the functional blocks shown in Figure 12.11.

The PSDsoft consists of seven modules: *PSDabel, PSD Configuration, PSD Fitter, PSD Simulator, Parallel Programmer, JTAG Programmer, and C Code Generator* [9].

The PSDabel accepts the standard known as HDL Abel environment, it provides an editor to create /edit a .abl file that can be used to define chip select logic, general-purpose logic, and PSD configuration parameters. When the .abl file is compiled, logic is synthesized and files are created and passed on to the PSDsoft fitting utility. The PSD Configuration is a utility used to specify the PSD MCU bus interface type, special I/O pin assignments, and particular internal PSD functions. The output of this module is the .glc configuration file, which is also used by the PSDsoft Fitter. The PSD Fitter has two main functions: the Fitter and the Address Translator. The Fitter accepts input from PSDabel and PSD Configuration, sythesizes this user logic and configuration, and fits the design to the PSD SILICON. The address Translator process allows the user to map the MCU firmware from a cross-compiler into the memory blocks within the PSD. As a result, the MCU firmware is merged with the logic and configuration definition of the PSD. The PSD Simulator is a WaferScale's version of SIMUCAD's SILOS simulation software, it provides functional chip-level simulation for PSD devices. PSDsoft automatically creates files for input to the simulator. These files convey relevant design information to the simulator. As a result, the user only has to create a stimulus file since all of the signals and node names are taken from the .abl file.

The PSD813F1 can be programmed by two means: in parallel form using the parallel programmer. It accepts the .obj file as input, allows viewing and editing of the .obj file, and programs and erases the PSD device. Through the JTAG Programmer which accepts the .obj file as input, and allows the PSD device to be programmed and erased in-system via the JTAG compatible Flash Link cable. The C code Generation module, is a new feature of PSDsoft that automatically generates C code functions and headers for controlling Flash PSD devices. These functions and headers are ANSI-C compatible. The generated files

(.c and .h) may be edited to suit the particular application, then compiled and linked with the rest of the code.

To realize the design first we have to create the design file with all the inputs and outputs, logical memory selection, etc. This is done using HDL language. For example, to specify the interface bus signals, we need to do the folliwing:

```
wr        pin;    "terminal 47, CNTL0"
rd        pin:    "terminal 50, CNTL1"
psen      pin;    "terminal 49, CNTL2"
ale       pin;    "terminal 10, PD0"
reset     pin;    "terminal 48, system reset"
a15..a0   pin;    "terminals 46..39, 37..30,
                  demultiplexed addresses"
```

As can be seen in this particular case it is not necessary to indicate the number of terminal to each signal, it is done automatically, for example the *wr* signal will be connected to terminal 47 known as CNTL0, by the manufacturer. It is the same for the *rd* signal to terminal 50 (CNTL1) and so on.

Next we can see in Figure 12.12 the programming of the Boolean equations in the PSD, where it is shown the three variables for the outputs f0, f1, f2,f3 assigned to pins 13, 14, 15, 16:

```
Title 'Look up table for the fuzzy controller
August 2001'.

"terminals assigned from table 4.
"Inputs
a,b,c,d,e,f  pin  2,3,4,5,6,7;

"Outputs
f3, f2, f1, f0 pin 12, 13, 14, 15 istype 'com';

bcd = [a,b,c];

ON. OFF =0, 1;

Truth_table (bcd -> [f3, f2, f1,f0])
1 ->  [0N,  OFF, OFF,  ON ];
2 ->  [ON,  OFF  OFF,  ON ];
3 ->  [ON,  OFF, OFF,  OFF];
4 ->  [ON,  OFF, OFF,  OFF];
5 ->  [OFF, OFF, OFF,  OFF];
6->   [ON,  OFF, OFF,  OFF];
7 ->  [OFF, ON,  ON,   ON];
8->   [0FF, ON,  ON,   OFF];
9->   [0FF, ON,  OFF,  ON];
10-> [0N,  OFF,  OFF,  OFF];
11-> [0N,  OFF,  OFF,  OFF];
12-> [0FF, ON,   ON,   ON];
1 3->[0FF, ON,   OFF,  OFF];
14 ->[OFF, ON,   OFF,  OFF];
15 ->[OFF, OFF,  OFF,  OFF];
16 ->[ON,  OFF,  OFF,  OFF];
17 ->[OFF, ON,   ON,   OFF];
18-> [OFF, ON,   OFF,  ON];
19 ->[OFF, ON,   OFF,  OFF];
20-> [0FF, OFF,  ON,   ON]
21-> [OFF, OFF,  OFF,  OFF];
22-> [OFF, ON,   OFF,  OFF];
23-> [OFF, OFF,  ON,   ON];
24-> [OFF, OFF,  ON,   ON];
25-> [OFF, OFF,  ON,   OFF];

" test_vectors edited
end
```

Figure 12.12 Programming the look up table in the PSD.

12.2.4 Experimental Results

In general the results of the fuzzy control on the CPLD, are almost the same as those obtained with the MATLAB fuzzy toolbox, as can be seen in Figure 12.13.

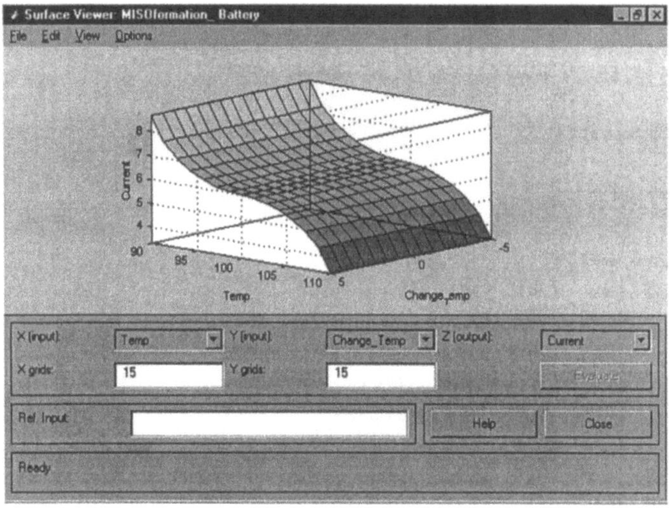

Figure 12.13 Temperature, ΔTemperature and Current curve.

Of course, there are some other options to implement in hardware a fuzzy controller, however here we have an advantage, a CPLD with the capacity required embedded in the PSD.

12.3 Automated Quality Control of Batteries

In this section we describe an embedded fuzzy control system for a quality control test in the manufacturing of power batteries. This battery test consists in discharging a battery for a certain time at a constant current value, all depending on the battery model. This system is able to improve on others, as it minimizes the problems presented when the test is initiated and when there are false contacts during the test, protecting the battery itself and test equipment. At the moment, it is used a proportional controller that even has demonstrated to be efficient, it has the disadvantages mentioned above.

In this fuzzy logic system for the quality control of batteries, we describe the use of a high scale integration device as the FLASH PSD813F1, which is a micro-controller that can support this implementation. It includes a CPLD (COMPLEX PLD) embedded, which allows implementing logical functions defined by the user in its 73 inputs 19 outputs AND-OR array. The fuzzy logic algorithm is summarized in a look up table, described by Boolean equations implemented in the PSD813F1 CPLD.

The fuzzy control system that has been implemented in the CPLD of the PSD813F1, basically controls the exciting current applied to the transistors bank to discharge the battery depending on its discharge current. This fuzzy system has two inputs, represented by the error and the change of the error, and one output, the current.

12.3.1 Introduction

In the manufacturing of batteries, one of the quality control tests consists in verifying if the battery has the electrical characteristics that corresponds to the model that it is testing, specifically in terms of the current (Hehner, 1981). This test basically consists in connecting an electronic load, represented in a bank of power transistors, controlled by a fuzzy control system, and discharge the battery for a certain time.

The battery and the electronic load, form a non-linear dynamic system that due to its characteristics, it is susceptible to be controlled through a fuzzy system (Klir & Yuan, 1995). In Figure 12.14 it is shown the basic control structure of this system. In this diagram the reference current corresponds to the discharge current for a certain model. In this case it is taken a battery model whose discharge current is 85 amperes, 6.43 volts as an initial voltage, 2.2 volts as final voltage and 2.5 seconds as the discharge time.

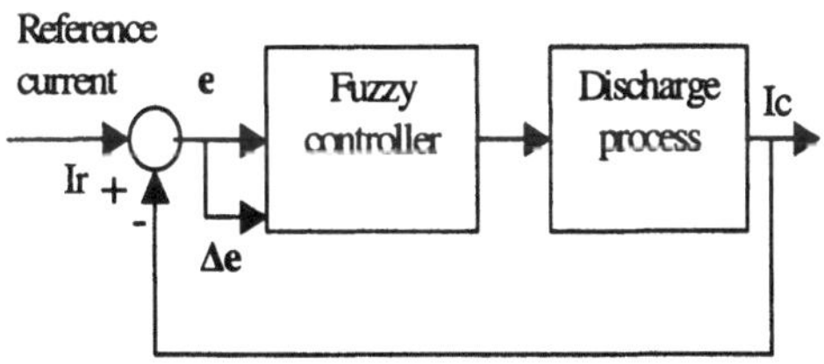

Figure 12.14 Fuzzy Control System.

The fuzzy controller will maintain a constant discharge current, 85 amperes in our case, inspite of the dynamic conditions of the battery because the voltage and the current are varying while the battery is being discharge.

The feedback of the system consists in reading the actual current and inform the controller, so this will be able to realize the fuzzy control. So this is a closed loop fuzzy control system.

In Figure 12.15, the fuzzy controller sends an exciting current to the Q_1 base, through the D/A converter that applies a V_{dac} voltage to R_{bexc}.

The principal purpose of Q_1 is to supply a higher base current to the transistors array in Q_2 so they can drain from the battery the I_C needed.

In reality Q_2 is an array of *n* power transistors connected in parallel. The exciting current I_{bexc} is given by,

$$I_{bexc} = V_{dac}/R_{bexc} \tag{12.2}$$

so that,

$$I_{exc} \approx \beta_1 I_{bexc} \tag{12.3}$$

If I_C is the discharge current of the battery, then the current needed to excite the load would be,

$$I_{bexc} = nI_C/(\beta_1\beta_2) \tag{12.4}$$

where β_1 and β_2 are the gains of the Q_1 transistors and the array of transistors Q_2 respectively, and they are non-linear. On the other hand, since I_C is also given by,

$$I_C=(V_{bat}-V_{CE})/R_C \tag{12.5}$$

and as the battery is in a discharge process, then V_{bat} diminishes in function of time, so we have a dynamic system than constantly demands adjustments in the base current of the array of transistors Q_2, and in consequence in the I_{bexc} of Q_1. This fact will require the D/A converter to continually adjust its output.

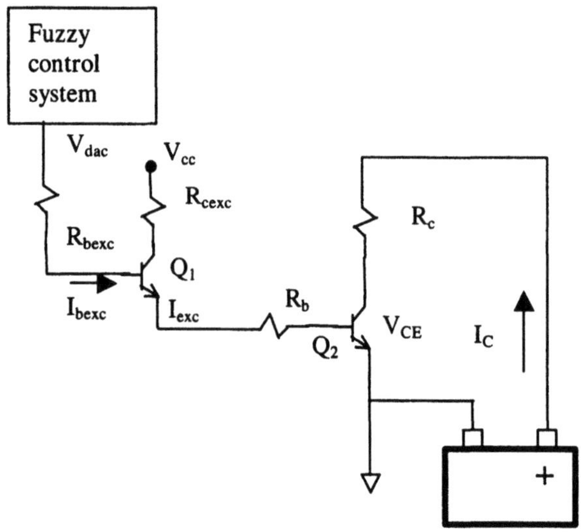

Figure 12.15 Fuzzy control diagram of the electronic charge.

In reference to the conditions mentioned above, the fuzzy control is a very good solution for this typc of problems.

In this work the fuzzy control system for quality control of batteries, that was implemented in the CPLD of the PSD813F1 (Harris & Rootz 1999), basically controls the exciting current applied to the transistors bank depending on the current in the battery. This fuzzy system has two inputs, represented by the error and the change of the error, and one output, the current.

12.3.2 Fuzzy Controller

The method described here to implement the fuzzy controller in the CPLD consists of logically synthesize the controller into a set of Boolean equations (Manzoul., 1994), then reduce it to a look up table. After the knowledge base and the set of all possible inputs are known, the look up table depends on the number of singletons of the input variables. Precisely the number of rows is equal to the product of the number of singletons of each input (Manzoul & Jayabharathi, 1991). The look up table is described by a set of Boolean equations which is an acceptable input to the CPLD.

To realize the fuzzy controller on a CPLD, first of all, we have to reduce the fuzzy controller to a set of Boolean equations and then these equations can be implemented on the CPLD.

For a two-input single-output fuzzy controller, the IF-THEN rule is written as:

$$\text{IF } x \text{ is } A_i \text{ AND } y \text{ is } B_i \text{ THEN } z \text{ is } C_i \tag{12.6}$$

Here x and A_i are defined in the universe of discourse A^I, y and B_i are defined in the input universe of discourse B^I, while z and C_i are defined in the output universe of discourse C^I .

This fuzzy rule can be put into thc simplcr form

$$\text{A x B} \rightarrow \text{C} \tag{12.7}$$

this rule can be transformed into a ternary fuzzy relation *Rm* based on Mamdani's fuzzy implication function (Jang, Sun & Mizutani, 1997):

$$\begin{aligned} R_m(A,B,C) &= (AxB)xC = \\ &= \int_{XxYxZ} \mu_A(x) \wedge \mu_B(x) \wedge \mu_C(x) / (x,y,z). \end{aligned} \tag{12.8}$$

where the resulting C' is expressed as

$$C' = (A' x B') \circ (A x B \rightarrow C). \qquad (12.9)$$

which can be decomposed as (Jang, Sun & Mizutani, 1997):

$$C' = [A' \circ (A \rightarrow C)] \cap [B' \circ (B \rightarrow C)] \qquad (12.10)$$

The preceding expression states that the resulting consequence C' can be expressed as the intersection of

$$C'_1 = A' \circ (A \rightarrow C) \qquad (12.11)$$

and

$$C'_2 = B' \circ (B \rightarrow C) \qquad (12.12)$$

each of one corresponds to a single-input single-output fuzzy rule as

$$\text{IF } \boldsymbol{x} \text{ is } \boldsymbol{A_i} \text{ THEN } \boldsymbol{y} \text{ is } \boldsymbol{B_i} \qquad (12.13)$$

where x and A_i are defined in the universe of discourse A^I, while y and B_i are defined in the output universe of discourse B^I.

In expression (12.13), if the dimensions of the two universes of discourse A^I and B^I are:

$$\dim[\boldsymbol{A^I}] = \boldsymbol{q} \qquad (12.14)$$
$$\dim[\boldsymbol{B^I}] = \boldsymbol{p} \qquad (12.15)$$

where $\boldsymbol{q}$ is the number of singletons in the input universe of discourse, and $\boldsymbol{p}$ is the number of singletons in the output universe of discourse, then the fuzzy relation of the system is computed as (Melin & Castillo, 2002),

$$\boldsymbol{R} = \vee (\boldsymbol{A_i}\ \boldsymbol{B_i}) \qquad (12.16)$$

where $\vee$ is the ***max*** operator, the dimension of the relation is:

$$\dim[\boldsymbol{R}^1] = \boldsymbol{q} \times \boldsymbol{p} \qquad (12.17)$$

So given any input $\boldsymbol{x}$, the fuzzy output $\boldsymbol{B}$ is computed using the compositional rule of inference;

$$\mathrm{y} = \boldsymbol{x} \circ \boldsymbol{R} \qquad (12.18)$$

where $\circ$ is the max-min composition of the fuzzy relation.

So for the two-input single output

$$C' = C'_1 \cap C'_2 \tag{12.19}$$

we have for C'_1

$$R'_1 = \vee (A_i \; C_i) \tag{12.20}$$

and for C'_2

$$R'_2 = \vee (B_i \; C_i) \tag{12.21}$$

For any inputs x, y, the fuzzy output C' is computed using the intersection of the compositional rules as:

$$C' = (x \circ R'_1) \cap (y \circ R'_2) \tag{12.22}$$

To achieve a good control in the test of the batteries we consider a fuzzy controller. The system can be very complex in a real situation, but for space reasons we describe it with the following linguistic variables:

input: error **e**, with three membership functions: Negative **Big NB**, Zero **ZERO** and **Positive Big PB**.

input: Change in error **Δe**, with three membership functions: Negative **Big NB**, Zero **ZERO** and **Positive Big PB**.

output: Current **C**, with three membership functions: Negative **Big NB**, Normal NORMAL and **Positive Big PB**.

Here error **e**, is the difference between the current reference I_r (85 amperes for this case) and the discharge current I_c, as can be seen in Figure 12.14; Current **C**, is the base current I_{bexc} of Q_1, according to Figure 12.15.

The system is described by the following three rules:

Rule 1: IF error is **NB** AND **Δe** is **NB**
THEN Current is **NB**

Rule 2: IF error is **ZERO** AND **Δe** is **ZERO**
THEN Current is NORMAL

Rule 3: IF error is **PB** AND **Δe** is **PB**
THEN Current is **PB**

The original ranges for the inputs **e** and Δ**e** and output are: -10 to 10 and –5 to 5 respectively , and 0.05 to .65 Amperes for the output current C. The membership values are in the interval [0, 1], where 0 denotes no membership and 1 denotes full membership. Figure 12.16 shows the membership functions of the linguistic variables.

Assume dim[*A*] = dim[*B*] = dim[*C*]= 5, and the three rules of the system are expressed numerically as:

Rule 1: IF [1.0 .5 0 0 0] AND [1.0 .6 0 0 0]
THEN [1.0 .5 0 0 0]

Rule 2: IF [0 .5 1.0 .5 0] AND [0 .3 1 .3 0]
THEN [0 .5 1.0 .5 0]

Rule 3: IF [0 0 0 .5 1.0] AND [0 0 0 .6 1]
THEN [0 0 0 .5 1.0]

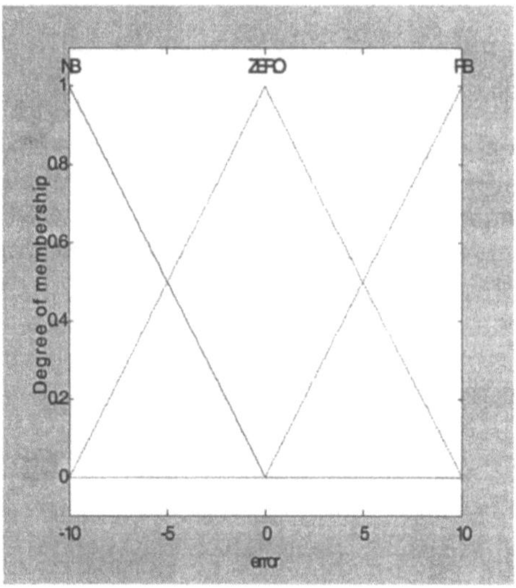

a)

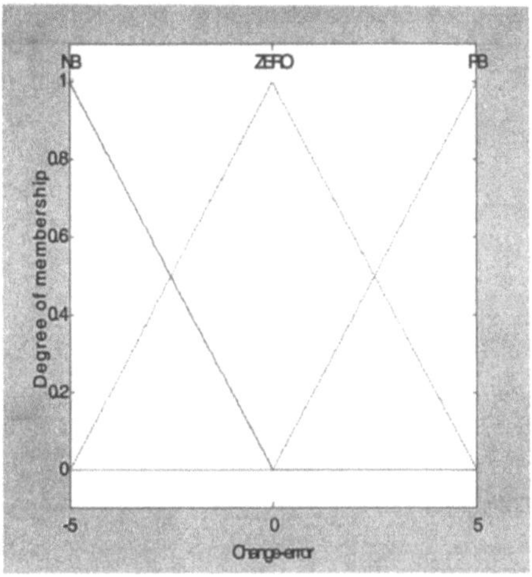

b)

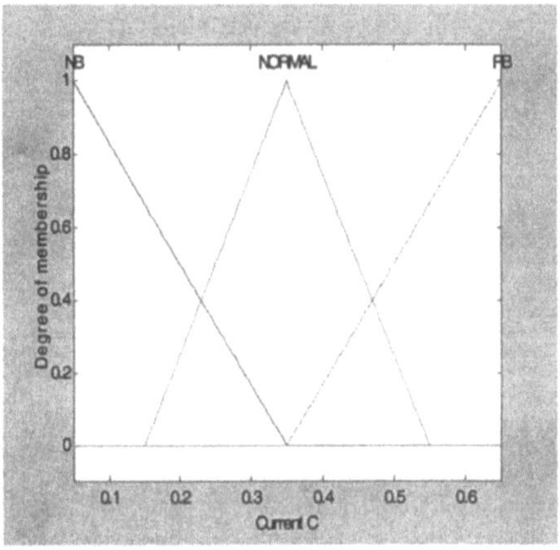

c)

Figure 12.16 Membership functions, a) input error, e., b) input change in error Δe, c) output Current C.

Using equation for the fuzzy relation $\boldsymbol{R'_1}$, we have:

$$R'_1 = \begin{bmatrix} 1 & .5 & 0 & 0 & 0 \\ .5 & .5 & .5 & .5 & 0 \\ 0 & .5 & 1 & .5 & 0 \\ 0 & .5 & .5 & .5 & .5 \\ 0 & 0 & 0 & .5 & 1 \end{bmatrix}$$

and for the fuzzy relation $\boldsymbol{R'_2}$,

$$R'_2 = \begin{bmatrix} 1 & .5 & 0 & 0 & 0 \\ .6 & .5 & .3 & .3 & 0 \\ 0 & .5 & 1 & .5 & 0 \\ 0 & .3 & .3 & .5 & .6 \\ 0 & 0 & 0 & .5 & 1 \end{bmatrix}$$

For example if the error is –4.13, and the change of the error is –0.81 that corresponds to the fuzzy values

-4.13 = [.0 1.0 .0 .0 .0]

and

-0.81 = [.0 .0 1.0 .0 .0]

as can be seen on Tables 12.7 and 12.8 respectively.

Table 12.7 Summary of the controller computations for A' ∘ (A→C).

	Input (error)	Fuzzified input	Fuzzified output
A1	∞ to –10	1.0 .0 .0 .0 .0	1.0 .5 .0 .0 .0
A2	-9.99 to –3.4	.0 1.0 .0 .0 .0	.5 .5 .5 .5 .0
A3	-3.3 to 3.4	.0 .0 1.0 .0 .0	.0 .5 1.0 .5 .0
A4	3.5 to 10	.0 .0 .0 1.0 .0	.0 ..5 .5 .5 5
A5	10 to ∞	.0 .0 .0 .0 1.0	.0 .0 .0 .5 1.0

Table 12.8 Summary of the controller computations for B' ∘ (B→C).

	Input (ΔTemp)	Fuzzified input	Fuzzified output
B1	0 to –5	1.0 .0 .0 .0 .0	1.0 .5 .0 .0 .0
B2	-4.9 to –1.7	.0 1.0 .0 .0 .0	.6 . 5 .3 .3 .0
B3	-1.6 to 1.7	.0 .0 1.0 .0 .0	.0 .5 1.0 .5 .0
B4	1.8 to 5	.0 .0 .0 1.0 .0	.0 .3 .3 .5 .6
B5	5.1 to ∞	.0 .0 .0 .0 1.0	.0 .0 .0 .5 1.0

To determine the fuzzy output, we employed the intersection of the compositional rules of inference given on row A2B3 of Table 12.9.

Table 12.9 Summary of the controller computations for: C'= [A' ∘ (A→C)] ∩ [B' ∘ (B→C)].

	Fuzzified output	Output .15(n-1)+.05
A1B1	1.0 .5 .0 .0 .0	1
A1B2	.6 .5 .0 .0 .0	1
A1B3	.0 .5 .0 .0 .0	2
A1B4	.0 .3 .0 .0 .0	2
A1B5	.0 .0 .0 .0 .0	5
A2B1	.5 .5 .0 .0 .0	1
A2B2	.5 .5 .3 .3 .0	2
A2B3	.0 .5 .5 .5 .0	3
A2B4	.0 .3 .3 .5 .0	3
A2B5	.0 .0 .0 .5 .0	4
A3B1	.0 .5 .0 .0 .0	2
A3B2	.0 .5 .3 .3 .0	2
A3B3	.0 .5 1.0 .5 .0	3
A3B4	.0 .3 .3 .5 .0	3
A3B5	.0 .0 .0 .5 .0	4
A4B1	.0 .5 .0 .0 .0	2
A4B2	.0 .5 .3 .3 .0	2
A4B3	.0 .5 .5 .5 .0	3
A4B4	.0 .3 .3 .5 .5	3
A4B5	.0 .0 .0 .5 .5	4
A5B1	.0 .0 .0 .0 .0	5
A5B2	.0 .0 .0 .3 .0	4
A5B3	.0 .0 .0 .5 .0	4
A5B4	.0 .0 .0 .5 .6	4
A5B5	.0 .0 .0 .5 1.0	5

The fuzzy output would be:

Fuzzy output = [.0 .5 .5 .5 .0]

so the output in the CPLD of the PSD813F1, is given by:

f2	f1	f0
0	1	1

as can be seen in Table 12.10, and equal to 3 in Table 12.9. This value is read by the 8031 microcontroller and then with the expression .15(n-1) + 0.05, where n is the value read in f2 f1 f0, we have that the current after deffuzzification is found to be, 0.35 Amps. In Table 12.9 we can see the complete computations of the controller, and the look up table representing the fuzzy controller is given in Table 12.10.

Table 12.10 Look up table for the fuzzy controller.

Input	Output f2 f1 f0
A1B1	0 0 1
A1B2	0 0 1
A1B3	0 1 0
A1B4	0 1 0
A1B5	1 0 1
A2B1	0 0 1
A2B2	0 1 0
A2B3	0 1 1
A2B4	0 1 1
A2B5	1 0 0
A3B1	0 1 0
A3B2	0 1 0
A3B3	0 1 1
A3B4	0 1 1
A3B5	1 0 0
A4B1	0 1 0
A4B2	0 1 0
A4B3	0 1 1
A4B4	0 1 1
A4B5	1 0 0
A5B1	1 0 1
A5B2	1 0 0
A5B3	1 0 0
A5B4	1 0 0
A5B5	1 0 1

Finally the look up table can be described by three variable Boolean equations:

$f0 = \Sigma\ (1,2,5,6,8,9,13,14,18,19,21,25)$

$f1 = \Sigma\ (3,4,7,8,9,11,12,13,14,16,17,18,19)$

$f2 = \Sigma\ (5,10,15,20,21,22,23,24,25)$

To compare the output results of the fuzzy system proposed for the CPLD, we implemented in the MATLAB fuzzy logic toolbox, a fuzzy system with two inputs and one output as can be seen in Figure 12.17. We have three membership functions for input error, NB (Negative Big) ZERO and PB (Positive Big); three membership functions for input Δerror, NB (Negative Big) ZERO and PB (Positive Big) as shown in Figure 12.16a and 12.16b. For the output Current, there are also three membership functions, NB (Negative Big) NORMAL and PB (Positive Big) as shown in figure 12.16c.

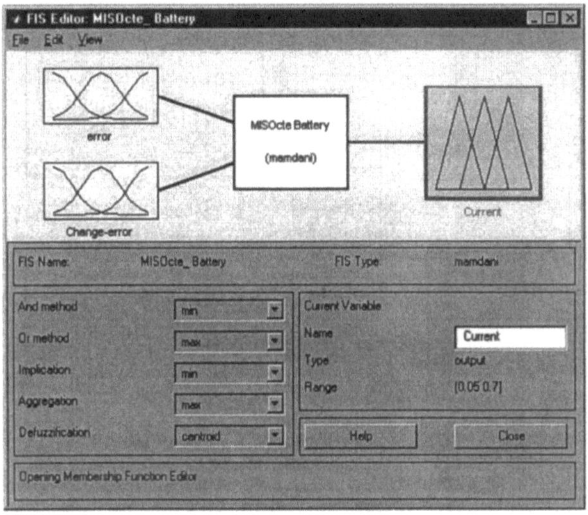

Figure 12.17 Two inputs, one output Fuzzy System.

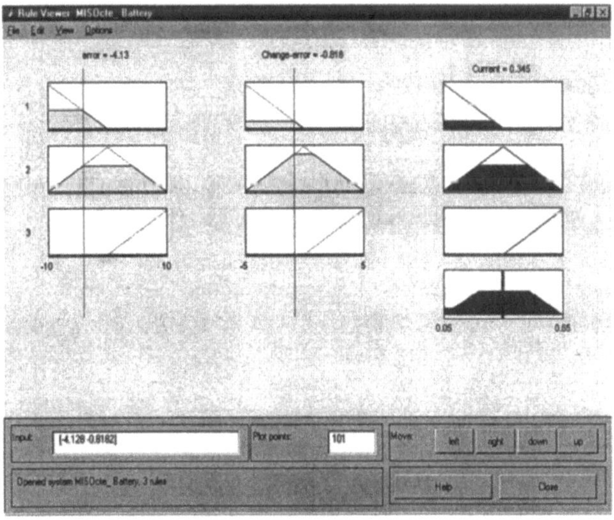

Figure 12.18 Summary of the fuzzy control for the inputs –4.13 and -.81, with an output=0.345 Amps.

In figure 12.18 we can see the output value 3.45, which corresponds to –4.13 as input error, in the range -9.99 to –3.4 of the table 1, and to –0.81 as input of change of error in the range –1.6 to 1.7 of table 12.8.

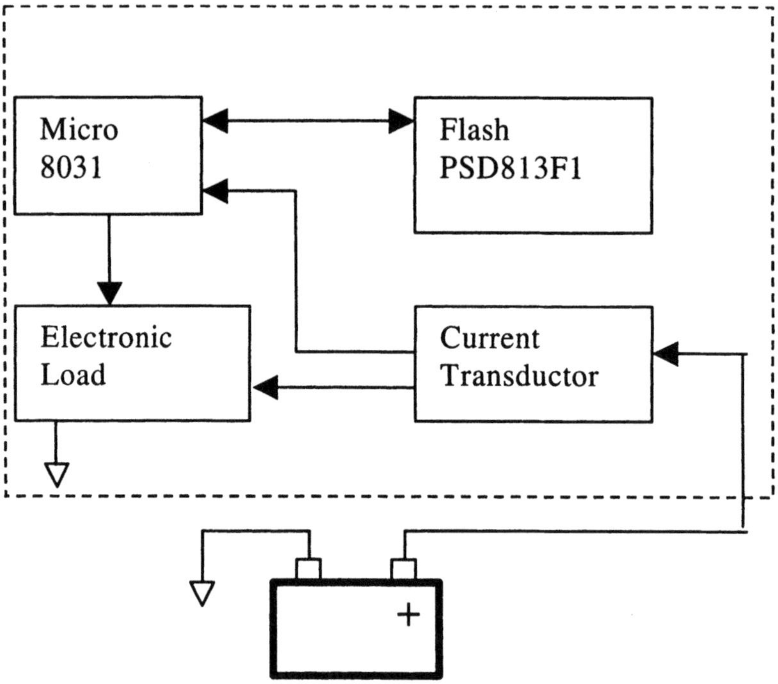

Figure 12.19 Automatic quality control test battery equipment.

12.3.3 Fuzzy Control Implementation

The Boolean equations obtained as a result of the logic synthesis of the fuzzy controller can be implemented in the CPLD of the PSD813F1. Figure 12.19 shows the diagram for the fuzzy control of the quality control test of batteries.

The PSD813F1 family includes a JTAG channel for In-System Programming (ISP). This ISP function is an extension of the typical JTAG boundary-scan function. It is an implementation of the JTAG-ISC (In-System Configuration) specification that is becoming an industry standard. The PSD allows the user to define and configure the JTAG ISC interface to meet the specifications, and needs no assistance from the embedded microcontroller (MCU), instead it uses the JTAG channel and the PSDsoft, which is an integrated system development software tool.

To program the fuzzy controller, it is necessary to configure and program the PSD, through the PSDsoft. This tool runs on a PC in the Windows 95/98 and Windows NT environments. PSDsoft supports the configuration of the functional blocks shown in Figure 12.19.

PSDsoft consists of seven modules: *PSDabel, PSD Configuration, PSD Fitter, PSD Simulator, Parallel Programmer, JTAG Programmer, and C Code Generator.*

The PSDabel accepts the standard known as HDL Abel environment, it provides an editor to create /edit a .abl file that can be used to define chip select logic, general-purpose logic, and PSD configuration parameters. When the .abl file is compiled, logic is synthesized and files are created and passed on to the PSDsoft fitting utility. The PSD Configuration is a utility used to specify the PSD MCU bus interface type, special I/O pin assignments, and particular internal PSD functions. The output of this module is the .glc configuration file, which is also used by the PSDsoft Fitter. The PSD Fitter has two main functions: the Fitter and the Address Translator. The Fitter accepts input from PSDabel and PSD Configuration, sythesizes this user logic and configuration, and fits the design to the PSD SILICON. The address Translator process allows the user to map the MCU firmware from a cross-compiler into the memory blocks within the PSD. As a result, the MCU firmware is merged with the logic and configuration definition of the PSD. The PSD Simulator is a WaferScale's version of SIMUCAD's SILOS simulation software, it provides functional chip-level simulation for PSD devices. PSDsoft automatically creates files for input to the simulator. These files convey relevant design information to the simulator. As a result, the user only has to create a stimulus file since all of the signals and node names are taken from the .abl file.

The PSD813F1 can be programmed by two means: in parallel form using the parallel programmer. It accepts the .obj file as input, allows viewing and editing of the .obj file, and programs and erases the PSD device. Through the JTAG Programmer which accepts the .obj file as input, and allows the PSD device to be programmed and erased in-system via the JTAG compatible Flash Link cable. The C code Generation module, is a new feature of PSDsoft that automatically generates C code functions and headers for controlling Flash PSD devices. These functions and headers are ANSI-C compatible. The generated files (.c and .h) may be edited to suit the particular application, then compiled and linked with the rest of the code.

```
Title 'Look up table for the fuzzy controller
Oct 2001'.

"terminals assigned from table 4.
"Inputs
a,b,c,d,e,f   pin  2,3,4,5,6,7;

"Outputs
f2, f1, f0  pin  13, 14, 15 istype 'com';

bcd = [a,b,c];

ON. OFF =0, 1;

Truth_table (bcd -> [ f2, f1,f0])
1 ->  [ OFF, OFF,  ON ];
2 ->  [ OFF  OFF,  ON ];
3 ->  [ OFF,  ON,  OFF];
4 ->  [ OFF,  ON,  OFF];
5 ->  [ ON,   OFF, ON ];
6->   [ OFF,  OFF, ON ];
7 ->  [ OFF,  ON,  OFF];
8->   [ OFF,  ON,  ON ];
9->   [ OFF,  ON,  ON ];
10->  [ ON ,  OFF, OFF];
11->  [ OFF,  ON , OFF];
12->  [ OFF,  ON,  OFF];
1 3-> [ OFF,  ON,  ON ];
14 -> [ OFF,  ON,  ON ];
15 -> [ ON,   OFF, OFF];
16 -> [ OFF,  ON,  OFF];
17 -> [ OFF,  ON,  OFF];
18->  [ OFF,  ON,  ON ];
19 -> [ OFF,  ON,  ON ];
20->  [ ON,   OFF, OFF]
21->  [ ON,   OFF, ON ];
22->  [ ON,  OFF,  OFF ];
23->  [ ON,  OFF,  OFF ];
24->  [ ON,  OFF,  OFF];
25->  [ ON,  OFF,  ON ];
"  test_vectors edited
end
```

Figure 12.20 Programming the look up table in the PSD.

To realize the design first we have to create the design file with all the inputs and outputs, logical memory selection, etc. This is done using the HDL language. For example, to specify the interface bus signals, we need to do the following:

```
wr        pin;    "terminal 47, CNTL0"
rd        pin:    "terminal 50, CNTL1"
psen      pin;    "terminal 49, CNTL2"
ale       pin;    "terminal 10, PD0"
reset     pin;    "terminal 48, system reset"
a15..a0   pin;    "terminals 46..39, 37..30,
                  demultiplexed addresses"
```

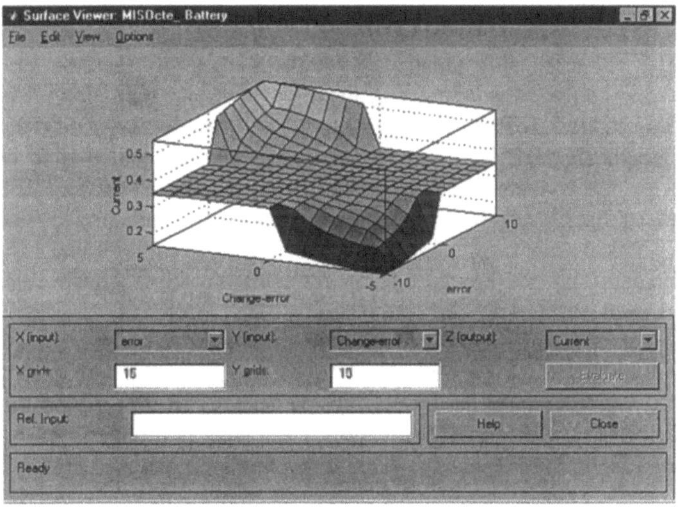

Figure 12.21 Error, Δerror and Current curve.

As can be seen in this particular case it is not necessary to indicate the number of terminal to each signal, this is done automatically, for example *wr* signal will be connected to terminal 47 known as CNTL0, by the manufacturer. It is the same for the *rd* signal to terminal 50 (CNTL1) and so on.

Next we can see in Figure 12.20 the programming of the Boolean equations in the PSD, where it is shown the three variables for the outputs f0, f1, f2 assigned to pins 13, 14, 15.

In general the results of the fuzzy control on the CPLD, are almost the same as those obtained with the MATLAB fuzzy logic toolbox, as can be seen in Figure 12.21. The high scale integration device, in this case the FLASH PSD813F1, allows us to implement fast and low cost fuzzy control systems. In this chapter for space reasons we only presented a two input one output fuzzy control with three membership functions each and 5 singletons as input, but it is also posible to implement more than that. Of course there are some other options to implement a hardware fuzzy controllers, however here we have an advantage, a CPLD with the capacity required embedded in the PSD.

12.4 Summary

We have described in this chapter different approaches for controlling an electrochemical process. We have shown that for this type of application the use of several soft computing techniques can help in reducing the time required to produce a battery. Even fuzzy control alone can reduce the formation time, but using neural networks and genetic algorithms reduces even more the time for production. Of course, this means that manufacturers can produce the batteries in half the time needed before. These are very good results for this application. A hardware implementation is still needed to completely automate the control of the process. Also, we consider that our hybrid approach can be used in a similar way for controlling similar non-linear dynamical processes for industrial applications.

The high scale integration device, such as the FLASH PSD813F1, allows us to implement fast and low cost fuzzy control systems. We have described the hardware implementation for controlling the charging process in the manufacturing of batteries. We have also described the hardware implementation of quality control test for the batteries after production. In this chapter, for space reasons we only presented hardware implementations of a two input one output fuzzy control (for both cases) with three membership functions each and five singletons as input, but it is also possible to implement more than that.

Bibliography

Abraham, E. & Firth, W. J. (1984). "Multiparameter Universal Route to Chaos in a Fabry-Perot Resonator", *Optical Bistability*, Vol. 2, pp. 119-126.

Albertos, P., Strietzel, R. & Mart, N. (1997). "Control Engineering Solutions: A Practical Approach", IEEE Computer Society Press.

Badiru, A.B. (1992). "Expert Systems Applications in Engineering and Manufacturing", Prentice-Hall.

Baker, J. E. (1987). "Reducing Bias and Inefficiency in the Selection Algorithms", Proceedings of the 2nd International Conference Genetic Algorithms, Hillsdale, NJ, pp. 14-21.

Barto, A. G., Sutton, R. S. & Anderson, C. (1983). "Neuron like Elements that can Solve Difficult Learning Control Problems", *IEEE Transactions on Systems, Man & Cybernetics*, Vol. 13, pp. 835-846.

Bernard, J. A. (1988). "Use of Rule-Based System for Process Control", IEEE Control Systems Magazine, Vol. 8, pp. 3-13.

Betin, F., Pinchon, D. & Capolino, G. A. (2000). "Fuzzy logic applied to speed control of a motor drive", *IEEE Transaction of Industrial Electronics*, Vol. 47, pp. 610-622.

Bezdek, J.C. (1981). "Pattern Recognition with Fuzzy Objective Function Algorithms", Plenum Press, New York.

Bode, H., Brodd, R.J. & Kordesch, K.V. (1977). Lead-Acid Batteries, John Wiley & Sons.

Bratko, I. (1990). "Prolog Programming for Artificial Intelligence", Addison Wesley.

Brock, W.A., Hsieh, D.A., & LeBaron, B. (1991). "Nonlinear Dynamics, Chaos and Instability", MIT Press, Cambridge, MA, USA.

Brogan, W. (1991). "Modern Control Theory", Prentice-Hall.

Bryson, A. E. & Ho Y.-C. (1969). "Applied Optimal Control", Blaisdell Press.

Cantú-Paz, E. (1995). "A Summary of Research on Parallel Genetic Algorithms", IlliGAL Report No. 95007, Illinois Genetic Algorithm Laboratory, University of Illinois at Urbana-Champaing.

Carpenter, G. A. & Grossberg, S. (1988). "The Art of Adaptive Pattern Recognition by a Self-Organizing Neural Network", IEEE Computer, Vol. 21 No. 3 pp. 77-88.

Castillo, O. & Melin, P. (1994a). "Developing a New Method for the Identification of Microorganisms for the Food Industry using the Fractal Dimension, *Journal of Fractals*, World Scientific, Vol. 2, No. 3, pp. 457-460.

Castillo, O. & Melin, P. (1994b). "An Intelligent System for Discovering Mathematical Models for Financial Time Series Prediction", Proceedings of TENCON'94, IEEE Computer Society Press, Vol. 1, pp. 217-221.

Castillo, O. & Melin, P. (1995a). "An Intelligent System for Financial Time Series Prediction Combining Dynamical Systems Theory, Fractal Theory and Statistical Methods", Proceedings of CIFER'95, IEEE Computer Society Press, pp.151-155.

Castillo, O. & Melin, P. (1995b). "Intelligent Model Discovery for Financial Time Series Prediction using Non-Linear Dynamical Systems and Statistical Methods", Proceedings of the Third International Conference on Artificial Intelligence Applications on Wall Street", Software Engineering Press, pp. 80-89.

Castillo, O. & Melin, P. (1995c). "An Intelligent System for the Simulation of Non-Linear Dynamical Economical Systems", *Journal of Mathematical Modelling and Simulation in Systems Analysis*, Edited by Achim Sydow, Gordon and Breach Publishers, Vol. 18-19, pp. 767-770.

Castillo, O. & Melin, P. (1996a). "Automated Mathematical Modelling for Financial Time Series Prediction using Fuzzy Logic, Dynamical Systems and Fractal Theory", Proceedings of CIFER'96, IEEE Computer Society Press, pp. 120-126.

Castillo, O. & Melin, P. (1996b). "Automated Mathematical Modelling and Simulation of Dynamical Engineering Systems using Artificial Intelligence Techniques", Proceedings CESA'96, Gerf EC Lille, pp. 682-687.

Castillo, O. & Melin, P. (1996c). "An Intelligent System for Financial Time Series Prediction using Fuzzy Logic Techniques and Fractal Theory", Proceedings ITHURS'96, Vol. 1, AMSE Press, pp. 423-430.

Castillo, O. & Melin, P. (1997a). "Mathematical Modelling and Simulation of Robotic Dynamic Systems using an Intelligent Tutoring System based on Fuzzy Logic and Fractal Theory", Proceedings of AIENG'97, Wessex Institute of Technology, pp. 97-100.

Castillo, O. & Melin, P. (1997b). "Mathematical Modelling and Simulation of Robotic Dynamic Systems using Fuzzy Logic Techniques and Fractal Theory", Proceedings of IMACS World Congress'97, Wissenschaft & Technik Verlag, Vol. 5, pp.343-348.

Castillo, O. & Melin, P. (1998a). "A New Fuzzy-Fractal-Genetic Method for Automated Mathematical Modelling and Simulation of Robotic Dynamic Systems", Proceedings of World Congress on Computational Intelligence FUZZ'98, IEEE Computer Society Press, Vol. 2, pp. 1182-1187.

Castillo, O. & Melin, P. (1998b). "Modelling, Simulation and Behavior Identification of Non-Linear Dynamical Systems with a New Fuzzy-

Fractal-Genetic Approach", Proceedings of IPMU'98, EDK Publishers, Vol. 1, pp. 467-474.

Castillo, O. & Melin, P. (1998c). "Modelling, Simulation and Forecasting of International Trade Dynamics using a New Fuzzy-Genetic Approach", Proceedings of CCM'98, AMSE Press, pp. 21-24.

Castillo, O. & Melin P. (1998d). "A New Fuzzy-Genetic Approach for the Simulation and Forecasting of International Trade Non-Linear Dynamics", Proceedings of CIFEr'98, IEEE Press, New York, NY, USA, pp. 189-196.

Castillo, O. & Melin, P. (1999a). "A New Fuzzy Inference System for Reasoning with Multiple Differential Equations for Modelling Complex Dynamical Systems", CIMCA'99, IOS Press, Vienna Austria, pp.224-229.

Castillo, O. & Melin, P. (1999b). "Intelligent Model-Based Adaptive Control of Robotic Dynamic Systems with a New Neuro-Fuzzy-Genetic Approach", Proceedings of Robotics and Applications, Acta Press, Santa Barbara, USA. pp. 270-275.

Castillo, O. & Melin, P. (1999c). "A General Method for Automated Simulation of Non-Linear Dynamical Systems using a New Fuzzy-Fractal-Genetic Approach", Proceedings CEC'99, IEEE Press, Vol. 3, pp. 2333-2340.

Castillo, O. & Melin, P. (1999d). "Automated Mathematical Modelling, Simulation and Behavior Identification of Robotic Dynamic Systems using a new Fuzzy-Fractal-Genetic Approach", *Journal of Robotics and Autonomous Systems*, Vol. 28, No. 1, pp. 19-30.

Castillo, O. & Melin, P. (1999e). "Automated Mathematical Modelling for Financial Time Series Prediction Combining Fuzzy Logic and Fractal Theory", Edited Book " Soft Computing for Financial Engineering", Springer-Verlag, Heidelberg, Germany, pp. 93-106.

Castillo, O. & Melin, P. (2000a). A New Method for Fuzzy Estimation of the Fractal Dimension and its Applications to Time Series Analysis and Pattern Recognition, Proceedings of NAFIPS'2000, Atlanta, GA, USA. pp. 451-455.

Castillo, O. & Melin, P. (2000b). "Intelligent Adaptive Model-Based Control of Robotic Dynamic Systems with a New Hybrid Neuro-Fuzzy-Fractal Approach", Proceedings of FLINS'2000, Belgium, World Scientific, pp. 351-358.

Castillo, O. & Melin, P. (2000c). "Modelling, Simulation and Behavior Identification of Non-Linear Dynamical Systems with a New Fuzzy-Fractal Genetic Approach", Edited Book in the Series of Advances in Fuzzy Systems edited by B. Bouchon, R. Yager, & L. A. Zadeh, World Scientific, pp. 95-106.

Castillo, O. & Melin, P. (2000d). "Hybrid Intelligent Systems for Time Series Prediction using Soft Computing Techniques and Fractal Theory", Proceedings of ANNIE'2000. St. Louis, Missouri, USA, AMSE Press, pp.1013-1018.

Castillo, O. & Melin, P. (2001a). "Soft Computing for Control of Non-Linear Dynamical Systems", Springer-Verlag, Heidelberg, Germany.

Castillo, O. & Melin, P. (2001b). "Intelligent Control of the Electrical Tuning Process in Televisions using Soft Computing Techniques", Proceedings of ANNIE'2001, St. Louis, Missouri, USA, pp. 209-214.

Castillo, O. & Melin, P. (2001c). "Soft Computing for Control of Non-Linear Dynamical Systems", Proceedings of the International Conference on Intelligent Technologies (InTech'2001), Bangkok, Thailand, pp. 1-7.

Castillo, O. & Melin, P. (2001d). "Hybrid Intelligent Systems for Time Series Prediction", *Journal of Smart Engineering System Design*, Francis and Taylor Publishers, Volume 3 Issue 3, pp. 1-8.

Castillo, O. & Melin, P. (2002). "A New Theory of Fuzzy Chaos and its Application for Simulation of Robotic Dynamic Systems", Proceedings of the 6^{Th} Joint Conference on Information Sciences, Durham, North Carolina, USA, pp. 149-152.

Chalmers, D. (1990). " The evolution of learning: an experiment in genetic connectionism". Proceedings of the 1990 Connectionist Models Summer School. Morgan Kaufmann.

Chen, G. & Pham, T. T. (2001). "Introduction to Fuzzy Sets, Fuzzy Logic, and Fuzzy Control Systems", CRC Press, Boca Raton, Florida, USA.

Chen, V. C. & Pao, Y. H. (1989). "Learning Control with Neural Networks", Proceedings of the International Conference on Robotics and Automation, pp. 1448-1453.

Chiang, L.H., Russell, E.L., & Braatz, R.D. (2000). "Fault Detection and Diagnosis in Industrial Systems", Springer-Verlag, Heidelberg, Germany.

Chiu, S., Chand, S., Moore, D. & Chaudhary, A. (1991). "Fuzzy Logic for Control of Roll and Moment for a Flexible Wing Aircraft", IEEE Control Systems Magazine, Vol. 11, pp. 42-48.

Chua, L.O. (1993). "Global unfolding of Chua's circuit", IEICE Transactions Fundam., pp. 704-734.

Cybenko, G. (1989). "Approximation by Superpositions of a Sigmoidal Function", *Mathematics of Control, Signals and Systems*, Vol. 2, pp. 303-314.

Davidor, Y. (1991). "Genetic Algorithms and Robotics: A Heuristic Strategy for Optimization", World Scientific Publishing.

DeJong, K. (1975). "The analysis and Behaviour of a Class of Genetic Adaptive Systems", PhD thesis, University of Michigan.

Denker, J. S. (1986). "Neural Networks for Computation", In J.S. Denker, editor, AIP Conference Proceedings, p. 151, American Institute of Physics.

Devaney, R. (1989). "An Introduction to Chaotic Dynamical Systems", Addison Wesley Publishing.

Dickason, V. (1997). The Loudspeaker Cookbook, McGraw Hill.

Du, R. (1998). "Engineering Monitoring and Diagnosis", Edited Book " Industrial and Manufacturing Systems", Academic Press, San Diego, California, USA, pp. 119-144.

Du, R., Elbestawi, M. A., & Wu, S.M. (1993). "Automated Monitoring of Manufacturing Processes", Transactions ASME J. Eng. Ind., pp. 245-252.

Fahlman, S. E. (1988). "Faster-leraning Variations on Back-propagation: an empirical study", Proceedings of the Conectionist Models Summer School, Carnegic Mellon University, pp. 38-51.

Fahlman, S. E. & Lebiere C. (1990). "The Cascade-Correlation Learning Architecture", Advances in Neural Information Processing Systems, Morgan Kaufmann.

Fu, K.S., Gonzalez, R.C. & Lee, C.S.G (1987). "Robotics: Control, Sensing, Vision and Intelligence", McGraw-Hill.

Geman, S. & Geman, D. (1984). "Stochastic Relaxation, Gibbs Distribution and the Bayesian Restoration in Images", *IEEE Transactions of Pattern Analysis and Machine Intelligence*, Vol. 6, pp. 721-741.

Gillies, A. M. (1985). "Machine Learning Procedures for Generating Image Domain Feature Detectors", PhD thesis, University of Michigan.

Goldberg, D.E. (1989). "Genetic Algorithms in Search, Optimization and Machine Learning", Addison Wesley Publishing.

Grebogi, C., Ott, E. & Yorke, J. A. (1987). "Chaos, Strange Attractors, and Fractal Basin Boundaries in Nonlinear Dynamics", Science, Vol. 238, pp. 632-637.

Gujarati, D. (1987). "Basic Econometrics", McGraw-Hill Publishing.

Gupta, M. M. & Sinha, N. K. (1996). "Intelligent Control Systems: Theory and Applications", IEEE Computer Society Press.

Hanselman D. & Littlefield B. (1995). "The Student Edition of MATLAB Version 4 : User's Guide", The Math-Works, Inc. Prentice-Hall.

Hehner, N. & Orsino, J.A. (1985). Storage Battery Manufacturing Manual III, Independent Battery Manufacturers Association.

Holland, J. H. (1975). "Adaptation in Natural and Artificial Systems", University of Michigan Press.

Hopfield, J. J. (1982). "Neural Networks and Physical Systems with Emergent Collective Computational Abilities", Proceedings of the National Academy of Science, USA, pp. 2554-2558.

Hollstien, R. B. (1971). "Artificial Genetic Adaptation in Computer Control Systems, PhD thesis, University of Michigan.

Hsu, C. (2000). "Advanced Signal Processing Technology by Soft Computing", World Scientific Press, Singapore.

Hunt, K. J., Sbarbaro, D., Zbikowski R. & Gawthrop, P. J. (1992). "Neural Networks for Control Systems-A survey", *Automatica*, Vol. 28 No. 6, pp. 1083-1112.

Ingber, L. & Rosen, B.E. (1992). "Genetic Algorithms and Very Fast Simulated Reannealing", *Journal of Mathematical and Computer Modelling*, Vol. 16, pp. 87-100.

Jamshidi, M. (1997). "Large-Scale Systems: Modelling, Control and Fuzzy Logic", Prentice-Hall.

Jain, A., Jain, A., Jain, S., & Jain, L. (2000). "Artificial Intelligence Techniques in Breast Cancer Diagnosis and Prognosis", World Scientific Press, Singapore.

Jang, J.-S. R. (1993). "ANFIS: Adaptive-Network-Based Fuzzy Inference Systems", *IEEE Transactions on Systems, Man and Cybernetics*, Vol. 23, pp. 665-685.

Jang, J.-S. R. & Gulley, N. (1997). "MATLAB: Fuzzy Logic Toolbox, User's Guide", The Math-Works, Inc. Publisher.

Jang, J.-S. R., Sun, C.-T. & Mizutani, E. (1997). "Neurofuzzy and Soft Computing: A Computational Approach to Learning and Machine Intelligence", Prentice-Hall.

Janikow, C. Z. & Michalewicz, Z. (1991). " An Experimental Comparison of Binary and Floating Point Representations in Genetic Algorithms", Proceedings of 4th International Conference Genetic Algorithms, pp. 31-36.

Kandel, A. (1992). "Fuzzy Expert Systems", CRC Press Inc.

Kapitaniak, T. (1996). "Controlling Chaos: Theoretical and Practical Methods in Non-Linear Dynamics", Academic Press.

Karnik, N.N & Mendel, J.M. (1998). "An Introduction to Type-2 Fuzzy Logic Systems", Technical Report, University of Southern California.

Kasai, Y. & Morimoto, Y (1988). "Electronically Controlled Continuously Variable Transmission", Proceedings of International Congress Transportation Electronics.

Kenjo, T. & Sugawara, A. (1994). Stepping Motors, Oxford Science Publications, London, Great Britain.

Kirkpatrick, S., Gelatt, C. D. & Vecchi, M. P. (1983). "Optimization by Simulated Annealing", Science, Vol. 220, pp. 671-680.

Kitano, H. (1990). "Designing neural networks using genetic algorithms with graph generation system", *Journal of Complex Systems*, Vol. 4, pp. 461-476.

Klir G. J. & Yuan, B (1995). "Fuzzy Sets and Fuzzy Logic" Theory and Applications, Prentice Hall, New York, NY, USA.

Kocarev, L. & Kapitaniak, T. (1995). "On an equivalence of chaotic attractors", *Journal of Physics A.*, Vol. 28, pp. 249-254.

Kohonen, T. (1982). "Self-Organized Formation of Topologically Correct Feature Maps, *Biological Cybernetics*, Vol. 66 pp. 59-69.

Kohonen, T. (1984). "Self-Organization and Associate Memory. Springer-Verlag, London.

Kohonen, T. (1984). "The Neural Phonetic Typewriter", *Computer*, Vol. 21 pp. 11-22.

Kohonen, T. (1989). "Self-Organization and Associate Memory. Springer-Verlag, London, 3 rd Edition.

Kohonen, T. (1990). "Improved Version of Learning Vector Quantization" Proceedings of the IJCNN'90 San Diego, CA, USA, Vol. 1 pp. 545-550.

Korn, G. A. (1995). "Neural Networks and Fuzzy Logic Control on Personal Computers and Workstations", MIT Press.

Kosko, B. (1992). "Neural Networks and Fuzzy Systems: A Dynamical Systems Approach to Machine Intelligence", Prentice-Hall.

Kosko, B. (1997). "Fuzzy Engineering", Prentice-Hall.

Lee, S. &Kil, R. M. (1991). "A Gaussian Potential Function Network with Hierarchically Self-Organizing Learning", *Neural Networks* Vol. 4 No. 2, pp. 207-224.

Li, Y. F. & Lau,C. C. (1989). "Developments of fuzzy algorithms for servo systems", IEEE Control System Magazine, vol. 9, pp. 65-72.

Lilly, K. W. (1993). "Efficient Dynamic Simulation of Robotic Mechanisms", Kluwer Academic Press.

Lim, S.Y., Hu, J. & Dawson, D.M. (1996). "An Output Feedback Controller for Trajectory Tracking of RLED Robots using Observed Backstepping Approach", Journal of Robotics and Automation, pp. 149-160.

Lippmann, R. P. (1987). "An Introduction to Computing with Neural Networks", IEEE Acoustics, Speech, and Signal Processing Magazine, Vol. 4, pp. 4-22.

Loctite Co. (1999). Speaker Assembly Adhesives Guide.

Lowe, D. (1989). "Adaptive Radial Basis Function Nonlinearities, and the Problem of Generalization, Proceedings of the First IEEE International Conference on Artificial Neural Networks, London, pp. 171-175.

Madan, R. (1993). "Chua's Circuit: Paradigm for Chaos", World Scientific, Singapore.

Mamdani, E. H. (1974). "Application of fuzzy algorithms for control of a simple dynamic plant", Proceedings of IEEE, Vol. 121, pp. 1585-1588.

Mamdani, E. H. & Assilian, S. (1975). "An Experiment in Linguistic Synthesis with a Fuzzy Logic Controller", *International Journal of Man-Machine Studies*, Vol. 7, pp. 1-13.

Man, K. F., Tang, K.S. & Kwong, S. (1999). "Genetic Algorithms". Springer-Verlag.

Mandelbrot, B. (1987). "The Fractal Geometry of Nature", W. H. Freeman and Company.

Mandelbrot, B. (1997). "Fractals and Scaling in Finance", Springer-Verlag, Heidelberg, Germany.

Manzoul M. A., & Jayabharathi D. (1991). "Implementation of Fuzzy Controllers Using Combinational Gate Arrays", Proceedings of NAFIPS'91 pp. 163-167.

Margaliot, M. & Langholz, G. (2000). "New Approaches to Fuzzy Modeling and Control", World Scientific Press, Singapore.

Masters, T. (1993). "Practical Neural Network recipe in C++", Academic Press, Inc.

Melin, P. & Castillo, O. (1996). "Modelling and Simulation for Bacteria Growth Control in the Food Industry using Artificial Intelligence", Proceedings of CESA'96, Gerf EC Lille, pp. 676-681.

Melin, P. & Castillo, O. (1997a). " An Adaptive Model-Based Neural Network Controller for Biochemical Reactors in the Food Industry", Proceedings of Control'97, Acta Press, pp.147-150.

Melin, P. & Castillo, O. (1997b). "Mathematical Modelling and Simulation of Bacteria Growth in the Time and Space Domains using Artificial Intelligence, Dynamical Systems and Fractal Theory", Proceedings of AMS'97, Acta Press, pp. 484-487.

Melin, P. & Castillo, O. (1997c). "An Adaptive Neural Network System for Bacteria Growth Control in the Food Industry using Mathematical Modelling and Simulation", Proceedings of IMACS World Congress'97, Wissenschaft & Technik Verlag, Vol 4 pp. 203-208.

Melin, P. & Castillo, O. (1997d). "Automated Mathematical Modelling and Simulation for Bacteria Growth Control in the Food Industry using Artificial Intelligence and Fractal Theory", *Journal of Systems Analysis, Modelling and Simulation*, Edited by Achim Sydow, Gordon and Breach Publishers, Vol. 29, pp. 189-206.

Melin, P. & Castillo, O. (1998a). "An Adaptive Model-Based Neuro-Fuzzy-Fractal Controller for Biochemical Reactors in the Food Industry", Proceedings of IJCNN'98, IEEE Computer Society Press, Vol. 1, pp.106-111.

Melin, P. & Castillo, O. (1998b). "A New Method for Adaptive Model-Based Neuro-Fuzzy-Fractal Control of Non-Linear Dynamic Plants: The Case of Biochemical Reactors", Proceedings of IPMU'98, EDK Publishers, Vol. 1, pp. 475-482.

Melin, P. & Castillo, O. (1999a). A New Method for Adaptive Model-Based Neuro-Fuzzy-Fractal Control of Non-Linear Dynamical Systems, Proceedings of the International Conference of Non-Linear Problems in Aviation and Aerospace'98, European Conference Publications, Daytona Beach, USA, pp. 499-506

Melin, P. & Castillo, O. (1999b). "A New Neuro-Fuzzy-Fractal Approach for Adaptive Model-Based Control of Non-Linear Dynamic Plants", Proceedings of Intelligent Systems and Control, Acta Press, Santa Barbara, USA. pp. 397-401.

Melin, P. & Castillo, O. (2000a). Controlling Complex Electrochemical Manufacturing Processes with Hybrid Intelligent Systems, Proceedings of NAFIPS'2000, Atlanta, GA, USA, pp. 490-494.

Melin, P. & Castillo, O. (2000b). "Controlling Chaotic and Unstable Behavior in Aircraft Dynamic Systems with a New Neuro-Fuzzy-Fractal Approach", Proceedings of IPMU'2000, Madrid, Spain, pp. 1791-1798.

Melin, P. & Castillo, O. (2000c). "Adaptive Intelligent Control of Aircraft Dynamic Systems with a new Hybrid Neuro-Fuzzy-Fractal Approach", Proceedings of FLINS'2000, Belgium, World Scientific, pp. 359-368.

Melin, P. & Castillo, O. (2000d). "Adaptive Model-Based Control of Non-linear Dynamical Systems with a Neuro-Fuzzy-Fractal Approach", Proceedings of ANNIE'2000. St. Louis Missouri, USA, AMSE Press, pp. 581-586.

Melin, P. & Castillo, O. (2001a). "A New Method for Adaptive Model-Based Control of Non-Linear Dynamic Plants using a Neuro-Fuzzy-Fractal Approach", *Journal of Soft Computing*, Springer-Verlag, 5(2), pp. 171-177.

Melin, P. & Castillo, O. (2001b) "Intelligent Control of Complex Electrochemical Systems With a Neuro-Fuzzy-Genetic Approach", *IEEE Transactions on Industrial Electronics*, Vol. 48 Number 5, pp. 951-955.

Melin, P. & Castillo, O. (2001c) "An Adaptive Model-Based Smart Controller of Bioreactors in the Food Industry Combining Neural Networks, Fuzzy Logic, and Fractal Theory", *Journal of Smart Engineering System Design*, Vol. 3, Number 1, pp. 65-79.

Melin, P. & Castillo, O. (2001d). "Controlling Electrochemical Processes using a New Neuro-Fuzzy-Genetic Approach", Proceedings of CBGIST'2001, Durham North Carolina, USA, pp. 159-163.

Melin, P. & Castillo, O. (2001e) "Adaptive Model-Based Control of Non-Linear Plants with a Neuro-Fuzzy-Genetic Approach", Proceedings of Artificial Intelligence and Soft Computing (ASC'2001), Cancun, Mexico, Acta Press, pp. 277-282.

Melin, P. & Castillo, O. (2001f) "Adaptive Control of Non-Linear Plants using Soft Computing and Mathematical Models", Proceedings of International Conference on Artificial Intelligence (IC-AI'2001), Las Vegas, USA, CSREA Press, pp. 1192-1198.

Melin, P. & Castillo, O. (2002) "Modelling Simulation and Control of Non-Linear Dynamical Systems: An Intelligent Approach Using Soft Computing and Fractal Theory", Taylor and Francis Publishers, London, Great Britain.

Mendel, J. (2001). "Uncertain Rule-Based Fuzzy Logic Systems: Introduction and New Directions", Prentice-Hall, New Jersey, USA.

Michalewicz, Z. (1996). "Genetic Algorithms + Data Structures = Evolution program", 3rd Edition, Springer-Verlag.

Miller, G., Todd,P. & Hedge, S. (1989). "Designing Neural Networks using Genetic Algorithms", Proceedings of the Third International Conference on Genetic Algorithms", Morgan Kauffmann.

Miller, W. T., Sutton, R. S. & Werbos P. J. (1995). "Neural Networks for Control", MIT Press.

Minsky, M. & Papert, S. (1969). "Preceptrons", MIT Press.

Mitchell, M. (1996). "An Introduction to Genetic Algorithms", MIT Press, Cambridge, MA, USA.

Montana, G. & Davis, L. (1989). "Training feedforward networks using genetic algorithms", Proceedings of the International Joint Conference on Artificial Intelligence, Morgan Kauffmann.

Moody, J. & Darken, C. (1989). "Fast Learning in Networks of Locally-Tuned Processing Units", *Neural Computation*, Vol. 1, pp281-294.

Moon, F.C. (1992). "Chaotic and Fractal Dynamics: An Introduction for Applied Scientists and Engineers", Wiley-Interscience, New York, USA.

Morari, M. & Zafiriou, E. (1989). "Robust Process Control", Prentice-Hall.

Moulet, M. (1992). "A Symbolic Algorithm for Computing Coefficients Accuracy in Regression", Proceedings of the International Workshop on Machine Learning, pp. 332-337.

Mühlenbein, H. & Schilierkamp, V. (1993). "Predictive Model for Breeder Genetic Algorithm". Evolutionary Computation. 1(1): 25-49.

Nakamura, S. (1997). "Numerical Analysis and Graphic Visualization with MATLAB", Prentice Hall.

Narendra, K. S. & Annaswamy, A. M. (1989). "Stable Adaptive Systems", Prentice Hall Publishing Company.

Ng, G. W. (1997). "Application of Neural Networks to Adaptive Control of Non-Linear Systems", John Wiley & Sons.

Ogorzalek, M. J. (1993). "Taming Chaos-Part II: Control" *IEEE Transactions on Circuit Systems*, Vol. 40, pp. 700-721.

Omidvar, O. & Elliot, D. L. (1997). "Neural Systems for Control", Academic Press, San Diego, CA, USA.

Ott, E., Grebogi, C. & Yorke, J.A. (1990). "Controlling Chaos". Physical Review Letters, Vol. 64, pp. 1196-1199.

Otten, R. H. J. M. & van Ginneken, L. P. (1989). "The Annealing Algorithm", Kluwer Academic.

Parker D. B. (1982). "Learning Logic", Invention Report S81-64, File 1, Office of Technology Licensing.

Patton, R.J., Frank, P.M., & Clark, R.N. (2000). "Issues of Fault Diagnosis for Dynamic Systems", Springer-Verlag, Heidelberg, Germany.

Peitgen, H.-O., Saupe, D., Jurgens, H. & Yunker, L. (1992). "Chaos and Fractals: New Frontiers of Science", Springer-Verlag, Heidelberg, Germany.

Pesin, Y. B. (1999). "Dimension Theory in Dynamical Systems: Contemporary Views and Applications", University of Chicago Press, USA.

Peters, E. (1994). "Fractal Market Analysis: Applying Chaos Theory to Investment and Economics", John Wiley & Sons, New York, USA.

Pham, D. T. & Xing, L. (1995). "Neural Networks for Identification, Prediction and Control", Springer-Verlag.

Pomerleau, D. A. (1991). "Efficient Training of Artificial Neural Networks for Autonomous Navigation", *Journal of Neural Computation*, Vol. 3, pp. 88-97.

Powell, M. J. D. (1987). "Radial Basis Functions for Multivariable Interpolation: a review", In J. C. Mason & M. G. Cox, editors, Algorithms for approximation, pp. 143-167, Oxford University Press.

Psaltis, D., Sideris, A. & Yamamura, A. (1988). "A Multilayered Neural Network Controller", IEEE Control Systems Magazine, Vol. 8, pp.17-21.

Pyragas, K. (1992). "Continuous control of chaos by self-controlling feedback", Physical Letters, pp. 421-428.

Rao, R. B. & Lu, S. (1993). "A Knowledge-Based Equation Discovery System for Engineering Domains", IEEE Expert, pp. 37-42.

Rasband, S.N. (1990). "Chaotic Dynamics of Non-Linear Systems", Wiley Interscience.

Rich, E. & Knight, K. (1991). "Artificial Intelligence", McGraw-Hill.

Ritter, H. & Schulten, K. (1987). "Extending Kohonen's Self-Organizing Mapping Algorithm to Learn Ballistic Movements", In Rolf Eckmiller & Christoph v.d. Malsburg, editors, Neural Computers, pp. 393-406, Springer-Verlag, London.

Rosenblatt, F. (1962). "Principles of Neurodynamics: Perceptrons and the Theory of Brain Mechanisms", Spartan.

Ruelle, D. (1990). "Deterministic Chaos: The Science and the Fiction", Proceedings of Roy. Soc. London, Vol. 427, pp. 241-248.

Rumelhart, D. E., Hinton, G. E. & Williams, R. J. (1986). "Learning Internal Representations by Error Propagation", Parallel Distributed Processing: Explorations in the Microestructure of Cognition, Vol. 1, Chapter 8, pp. 318-362, MIT Press, Cambridge, MA, USA.

Rumelhart, D. E., & Zipser, D. (1986). "Feature Discovery by Competitive Learning", In D. E. Rumelhart and James L. McClelland, editors, Parallel Distributed Processing: explorations in the microstructure of cognition, Vol. 1, Chapter 5, pp. 151-193, MIT Press, Cambridge, MA, USA.

Runkler, T. A. & Glesner, M. (1994). "Defuzzification and Ranking in the Context of Membership Value Semantics, Rule Modality, and Measurement Theory", Proceedings of European Congress on Fuzzy and Intelligent Technologies.

Ruspini, E. H. (1982). " Recent Development in Fuzzy Clustering", Fuzzy Set and Possibility Theory, North Holland, pp. 133-147.

Russell, S. & Norvig, P. (1995). "Artificial Intelligence: A Modern Approach", Prentice-Hall.

Russell, E.L., Chiang, L.H., & Braatz, R.D. (2000). "Data-driven Methods for Fault Detection and Diagnosis in Chemical Processes", Springer-Verlag, Heidelberg, Germany.

Sackinger, E., Boser, B. E., Bromley, J., LeCun, Y. & Jackel, L. D. (1992). "Application of the Anna Neural Network Chip to High-Speed Character Recognition", IEEE Transactions on Neural Networks, Vol. 3, pp. 498-505.

Saez, D. & Cipriano, A. (2001). "Optimization of Industrial Processes at Supervisory Level", Springer-Verlag, Heidelberg, Germany.

Samad, T. & Foslien, W. (1994). "Neural Networks as Generic Nonlinear Controllers", Proceedings of the World Congress on Neural Networks, pp. 191-194.

Scharf, L. L. (1991). "Statistical Signal Processing, Detection, Estimation, and Time Series Analysis", Addison-Wesley, Reading, MA, USA.

Sejnowski, T. J. & Rosenberg, C. R. (1987). "Parallel Networks that Learn to Pronounce English Text", *Journal of Complex Systems*, Vol. 1, pp. 145-168.

Semmes, S. (2000). "Some Novel Types of Fractal Geometry", Oxford University Press, England.

Sepulveda, R., Castillo, O., Montiel, O. and Lopez, M. (1998). "Analysis of Fuzzy Control System for the Process of Forming Batteries", Proceedings of ISRA'98, Coahuila, Mexico, pp. 203-210.

Sleeman, D. & Edwards, P. (1992). "Proceedings of the International Workshop on Machine Learning", Morgan Kauffman Publishers.

Soucek, B. (1991). "Neural and Intelligent Systems Integration: Fifth and Sixth Generation Integrated Reasoning Information Systems", John Wiley and Sons.

Spears, W. M. & DeJong, K. (1991). " An Analysis of Multi-point Crossover", In G. J. E Rawlins, editor Foundations of Genetic Algorithms, pp. 301-315.

Staib, W. E. (1993). "The Intelligent Arc Furance: Neural Networks Revolutionize Steelmaking", Proceedings of the World Congress on Neural Networks, pp. 466-469.

Staib, W. E. & Staib, R. B. (1992). "The Intelligent Arc Furnace Controller: A Neural Network Electrode Position Optimization System for the Electric Arc Furnace", Proceedings of the International Conference on Neural Networks, Vol. 3, pp. 1-9.

Su, H. T. & McAvoy, T. J. (1993). "Neural Model Predictive Models of Nonlinear Chemical Processes", Proceedings of the 8th International Symposium on Intelligent Control, pp. 358-363.

Su, H. T., McAvoy, T. J. & Werbos, P. J. (1992). "Long-term Predictions of Chemical Processes using Recurrent Neural Networks: A Parallel Training Approach", Industrial & Engineering Chemistry Research, Vol. 31, pp. 1338-1352.

Sueda, N. & Iwamasa, M. (1995). "A Pilot System for Plant Control using Model-Based Reasoning", IEEE Expert, Vol. 10, No. 4, pp.24-31.

Sugeno, M. & Kang, G. T. (1988). "Structure Identification of Fuzzy Model", *Journal of Fuzzy Sets and Systems*, Vol. 28, pp. 15-33.

Szu, H. & Hartley, R. (1987). "Fast Simulated Annealing", Physics Letters, Vol. 122, pp. 157-162.

Takagi, T. & Sugeno, M. (1985). "Fuzzy Identification of Systems and its Applications to Modeling and Control", *IEEE Transactions on Systems, Man and Cybernetics,* Vol. 15, pp. 116-132.

Tricot, C. (1995). "Curves and Fractal Dimension", Springer-Verlag, Heidelberg, Germany.

Troudet, T. (1991). "Towards Practical Design using Neural Computation", Proceedings of the International Conference on Neural Networks, Vol. 2, pp. 675-681.

Tsoukalas, L. H. & Uhrig, R. E. (1997). "Fuzzy and Neural Approaches in Engineering", John Wiley & Sons, Inc., New Jersey, USA.

Tsukamoto, Y. (1979). "An Approach to Fuzzy Reasoning Method", In Gupta, M. M., Ragade, R. K. and Yager, R. R., editors, Advanced in Fuzzy Set Theory and Applications, pp. 137-149, North-Holland.

Ungar, L. H. (1995). "A Bioreactor Benchmark for Adaptive Network-Based Process Control", Neural Networks for Control, MIT Press, pp. 387-402.

Von Altrock, C. (1995). "Fuzzy Logic & Neuro Fuzzy Applications Explained", Prentice Hall.

Wagenknecht, M. & Hartmann, K. (1988). "Application of Fuzzy Sets of Type 2 to the solution of Fuzzy Equation Systems", *Fuzzy Sets and Systems*, Vol. 25, pp. 183-190.

Wang, L.-X. (1997). "A Course in Fuzzy Systems and Control", Prentice-Hall, Upper Saddle River, NJ, USA.

Weigend, A. & Gershenfeld, N.A. (1994). "Time Series Prediction: Forecasting the Future and Understanding the Past", Addison Wesley Publishing.

Werbos, P. J. (1991). "An Overview of Neural Networks for Control", IEEE Control Systems Magazine, Vol. 11, pp. 40-41.

Werbos, P. J. (1974). "Beyond Regression: New Tools for Prediction and Analysis in the Behavioral Sciences", Ph.D. Thesis, Harvard University.

Widrow, B. & Stearns, D. (1985). "Adaptive Signal Processing", Prentice-Hall.

White, H. (1989). "An Additional Hidden Unit Test for Neglected Non-linearity in Multilayer Feedforward Networks", Proceedings of IJCNN'89, Washington, D.C., IEEE Press, pp. 451-455.

Wright, A. H. (1991). "Genetic Algorithms for Real Parameter Optimization", In J. E. Rawlins, editor, Foundations of Genetic Algorithms, Morgan Kaufmann, pp205-218.

Yager, R.R. (1980). "Fuzzy Subsets of Type II in Decisions", *Journal of Cybernetics*, Vol. 10, pp. 137-159.

Yager, R.R. & Filev, D.P. (1993). "SLIDE: A Simple Adaptive Defuzzification Method", *IEEE Transactions on Fuzzy Systems*, Vol. 1, pp. 69-78.

Yager, R. R. & Filev, D.P. (1994). "Generation of Fuzzy Rules by Mountain Clustering", *Journal of Intelligent and Fuzzy Systems*, Vol. 2, No. 3, pp. 209-219.

Yamamoto, Y. & Yun, X. (1997). "A Modular Approach to Dynamic Modelling of a Class of Mobil Manipulators", *Journal of Robotics and Automation*. pp. 41-48.

Yang, C. C., Marefat, M. M., & Ciarallo, F. W. (1999). "Modeling Errors for Dimensional Inspection using Active Vision", *Journal of Robotics and Computer-Integrated Manufacturing*, Pergamon Press, 15, pp. 23-27.

Yasunobu, S. & Miyamoto, S. (1985). "Automatic Train Operation by Predictive Fuzzy Control", Industrial Applications of Fuzzy Control, pp.1-18, North Holland.

Yen, J., Langari, R. & Zadeh, L. A. (1995). "Industrial Applications of Fuzzy Control and Intelligent Systems", IEEE Computer Society Press.

Yen, J. & Langari, R. (1999). "Fuzzy Logic: Intelligence, Control, and Information", Prentice Hall, New Jersey, USA.

Zadeh, L.A. (1965). "Fuzzy Sets", Journal of Information and Control, Vol. 8, pp. 338-353.

Zadeh, L.A. (1971a). "Similarity Relations and Fuzzy Ordering", Journal of Information Sciences", Vol. 3, pp. 177-206.

Zadeh, L. A. (1971b). "Quantitative Fuzzy Semantics", Journal of Information Sciences, 3, pp. 159-176.

Zadeh, L.A. (1973). "Outline of a New Approach to the Analysis of Complex Systems and Decision Processes", IEEE Transactions on Systems, Man and Cybernetics, Vol. 3, pp. 28-44.

Zadeh, L.A. (1975). "The Concept of a Linguistic Variable and its Application to Approximate Reasoning –1", Journal of Information Sciences, Vol. 8, pp. 199-249.

Zomaya, A. Y. (1992). "Modelling and Simulation of Robot Manipulators: A Parallel Processing Approach", World Scientific Publishing.

Index

MIX
Papier aus verantwortungsvollen Quellen
Paper from responsible sources
FSC® C105338

If you have any concerns about our products,
you can contact us on
ProductSafety@springernature.com

In case Publisher is established outside the EU,
the EU authorized representative is:
Springer Nature Customer Service Center GmbH
Europaplatz 3, 69115 Heidelberg, Germany

Printed by Libri Plureos GmbH
in Hamburg, Germany